Invisible Astronomy

Invisible Astronomy

Colin A. Ronan

J. B. LIPPINCOTT COMPANY

PHILADELPHIA & NEW YORK

1972

Foreword

by Fred Hoyle, F.R.S.

Plumian Professor of Astronomy and Experimental Philosophy and Director of Institute of Theoretical Astronomy, University of Cambridge.

Astronomy has had three phases of development, a classical phase, an astrophysical phase and now – very recently – the beginnings of a cosmological phase. Because today, with new vistas opening up, astronomy is in a fertile state, this book appears at a timely moment.

In its classical phase astronomy was the forerunner of the sciences. Perhaps the most important aspect of the early work lay in the demonstration that cause and effect can be described mathematically. This concept underlies the whole of modern science.

By the middle of the nineteenth century this early phase of development was essentially completed, The years from about 1850 to 1920, although by no means unimportant, were comparatively fallow. By way of contrast this was the great age of physics. Thermodynamics, classical electromagnetic theory, relativity, Planck, Bohr, and if we extend by five years or so, quantum mechanics, all belong to this tremendous age of physics.

The year 1920 forms a dividing line in astronomy because this was the approximate date of two developments of enormous significance, Hubble's discovery of the expanding universe and Eddington's early researches in stellar structure. The latter is more straightforward and I will take it first. Eddington surmised that the energy of the stars is produced by the fusion of hydrogen to helium. Within a few more years the theory of radiation and of ionized gases had been applied to the interiors of stars. It was found possible to work out fairly accurate values of stellar luminosities without a precise knowledge of nuclear physics being necessary, although by the 1930s the nuclear problem itself was solved quantitatively for the proton chain and for the CN cycle by Bethe in their simplest forms. This permitted stellar radii to be calculated as well as luminosities, and a beginning could be made to

v

solving the problem that had eluded astronomers for so long – the problem of stellar evolution. This problem has been the main topic, and its solution the main achievement, of astronomy over the past thirty years.

In the 1950s it was realized that nuclear reactions in stars can be viewed in two ways. If one is interested mainly in the structure of stars then the reactions have importance as an energy source but the products of the reactions are not of especial relevance, at any rate in detail. On the other hand one can turn the whole problem around and look on the stars as devices for changing the nuclear structure of matter. This is the important concept of nucleogenesis – that matter has a history which can be understood in detail, that we can explain the abundances of the chemical elements in terms of processes that take place within the stars. A decade or more of work on this idea has shown that it is essentially correct. Although it remains possible that an appeal to non-stellar processes may have to be made for a few special elements we now see that the outstanding features of the history of matter are to be understood in the stars. The working out of the complex details has become a mainstream of progress in astronomy. This is the astrophysical phase which I mentioned at the outset.

Now I come to the latest phase which began with Hubble's discovery of the expanding universe and which received great impetus in the 1950s with the development of radioastronomy. I do not think even the most enthusiastic radioastronomer could have guessed in 1950 just where his subject was going to lead. At first we all thought that the emission of radio waves by cosmic objects would be limited to stars not much different from the Sun and to hot clouds of gas. This point of view was not only wrong but fantastically wrong. I well remember the day when it seemed almost incredible that any of the strong radio sources could be extragalactic. Now we contemplate upwards of a million such sources and we place most of them at least as far away as the most distant galaxies that can be observed by optical means.

Quasars and radio galaxies belong to the same family of phenomena – both are radio sources. The new feature revealed by the quasars has been the variation of both radio output and optical output in short time periods, months or even days. This is scarcely understandable unless the emitting volumes are small. The same property is now being uncovered in the case of the extreme nuclei of radio galaxies. These it seems must be exceedingly small by galactic standards, not more than a light year in diameter, and perhaps even substantially less than this.

Furthermore, unprecedentedly large energies are involved, which demands that the objects in question have large masses, much greater than the masses of individual stars. The masses seem likely to be of a galactic order, so that we have the astonishing picture of many millions of solar masses confined to volumes not a great deal larger than the solar system, and emitting energy on a galactic scale – indeed emitting such a flood of energy that they can be observed even at very great distances.

As well as radio waves and optical light these same objects have recently been found to emit strongly in the infra-red. They probably are also active in the ultraviolet and X-ray regions of the spectrum but this is harder to verify by observation. In short we have a new class of object active over the whole spectrum, emitting radiation not by atomic transitions as in the case of stars but through other processes, in particular through synchrotron radiation.

It is safe to predict that the study of these objects will form a new major stream of astronomical development and that it will occupy much of the attention of astronomers over the remaining years of the present century. It is certain that we shall be led into strange territory, probably into the illusive realm of strong gravitational fields. And we can now contemplate an understanding of the evolution of galaxies just as the astronomers of thirty years ago could contemplate an eventual under-standing of the evolution of stars. Colin Ronan has hit on a good description of this third phase with his title *Invisible Astronomy*.

FRED HOYLE

24 June 1969

Contents

List of Plates

List of Figures

Acknowledgements

Acknowledgements and thanks for permission to reproduce photographs are due to Mt Wilson and Palomar Observatories for plates I, II, IIIa, VIa, and VIb; to the Lick Observatory for plates IIIb, VIc, and XII; to the Ronan Picture Library for plates IVa, IVb, and IVd; to the United States Information Service for plates IVc, VIIa, IX, and XI; to Professor G. P. Kuiper for plate Va; to Mr Gordon H. Pettengill for plate Vb; to Cornell University for plate VIIb; to the Royal Society for plate VIII; to the Royal Greenwich Observatory, Herstmonceux, for plates Xa and Xb; and to Culham Laboratory for plates Xc and Xd.

Preface

Astronomy is in an age of revolution: not since the first use of the telescope three and a half centuries ago has there been so extensive a change in our ideas. As in the case of the advent of the telescope, this change is due to completely new ways of observing the heavens, but now utilizing every possible advance in twentieth century technology. Rocket launchers, artificial satellites and, above all, electronic techniques, have made it possible to extend the range of observation well beyond the capabilities of the optical telescope. Visible light is no longer the only source of information – the astronomer can now probe in the infra-red and radio extensions at one end of the spectrum, and in the shorter ultra-violet, X-ray and gamma ray wavelengths at the other. In addition he is extracting evidence from new investigations with cosmic rays and other nuclear particles. The whole of observational astronomy has broadened in a way that was inconceivable two decades ago.

This is not to imply that visual observation has nothing more to contribute – the opposite is in fact true. The optical telescope has much to give in its own right, as well as in helping to understand and interpret the astonishing results that are arising from investigations in the invisible wavelengths of the spectrum. The scheme adopted in this book is therefore to begin by considering the role of observation and its relationship to theory – so important when results are to be interpreted, and when concepts with the entire universe as their canvas are to be assessed. Also, before discussing the achievements in invisible wavelengths, first to give a picture of the universe as seen by the visual astronomer alone, and the techniques and interpretations he has used in drawing his picture, and then to provide some background on the physics involved in the new investigations. Only in this way does it seem possible to attempt to present a coherent and intelligible survey of the new, invisible astronomy, and glance at some of its profound implications.

In preparing the book I have been helped by my wife, who has read it through and ironed out many inelegancies in the text, and my thanks are also due to Mrs E. Beebe for finding her way through many corrections and preparing a legible typescript. Lastly I should like to compliment Lawrence Clark on his admirable figures, and thank Mr Andrew Pennycook and Mr Patrick Moore for reading through the manuscript and the proofs.

Cowlinge,   COLIN A. RONAN
Suffolk,
January 1968

Since preparing the manuscript of the book, some recent observations have brought new evidence to bear on the problems discussed here, and where possible I have incorporated them in the text.

Cowlinge,   COLIN A. RONAN
Suffolk,
June 1969

CHAPTER 1

Introduction

Theory and Observation

From the time of primitive man the heavens have exerted a deep fascination. Sometimes they have appealed more to the poetic side of human nature, sometimes more emphatically to the religious and, most often, to the superstitious. Yet whichever aspect might be in the ascendant, there has always been a fourth – the philosophical – that has run beside the others like a thread, unbroken, even if, at times, it has been a little knotted. This philosophical or, as we should now say, scientific, aspect of the skies has changed considerably with the passing of time; over the years it has gradually become emancipated from the shackles of divine revelation and mystical influence, until it now stands alone to provide a factual picture of the universe that is astounding in its breadth. All the same, unanswered questions still make us realize that our knowledge is far from complete, and this incompleteness is far from superficial – it extends to some of the most basic problems, such as the birth and death of stars, planets, and the general life history of the universe itself. But this is bound to be, for the more we discover, the greater the host of problems that confronts us; as our techniques of observation develop, so entirely new questions present themselves. In fact, the ability we at any time possess to probe space colours and conditions our entire outlook.

To appreciate the importance of observation is vital if the significance of the new techniques of 'invisible' astronomy is to be realized, and the most satisfactory way to do this is probably to glance back at the changing picture that previous technical advances have brought in their train. For a great part of the history of astronomy, the technical aids that were available were few and simple. They were concerned with the measurement of position of one star relative to another, and even so there was a limit to the accuracy of which they were capable since they possessed no optical parts. A refined form of sight not dissimilar to that on a rifle represented the peak of their evolution, and since the average human eye cannot resolve as separate objects that appear nearer

1

together than about 3 minutes of arc* (3′) there is obviously a limit to the accuracy attainable with non-optical sights. In practice such observations of position were concerned with objects separated by more than this, and repeated measurements made it possible to achieve an accuracy of some 2′. Yet with this degree of precision it was not possible to determine stellar distances, and the lack of optical aid rendered any factual picture of the physical nature of the Moon and the planets impossible to attain.

The first recorded use of optical aids (telescopes) in astronomy occurred in 1609, and from thenceforward the whole emphasis began to change. Mountains and valleys could be seen on the Moon, some of the physical details of the nearer planets could be determined, it was discovered that Jupiter possessed satellites and, equally important, that the hazy band of light stretching across the heavens and known as the Milky Way was a concourse of myriads of separate stars. The Sun was found to have spots on its surface and, with the solution of the theoretical problem posed by the motions of the planets, observational astronomy centred on the solar system. Navigation, with its demand for the precise determination of longitude at sea, also provided an impetus to studies of the Moon's motion and it was to be many years before a concentration on stellar astronomy returned.

The advent of the telescope also brought about an increase in positional accuracy – a vital matter on the formulation of a precise planetary and gravitational theory. The 2′ of pre-telescopic days soon gave place to errors measured in seconds of arc, for even a very small telescope can increase the resolving power of the observer some fifty times. Consider, for instance, a simple terrestrial observation of two lighted windows. If these are six feet apart then they will be seen as separate by an observer with normal eyesight at a distance of four miles, but if they are any closer than six feet, they will seem to the observer to run into one another. Yet a small pair of binoculars will allow one to pick out separate panes, and telescopes were soon developed that would allow an observer to detect minute differences in the curtain material. As time went on and larger and more efficient telescopes were constructed,

* For those unfamiliar with minutes and seconds of arc, it may be as well to give their relationships here. The term 'of arc' is usually employed to distinguish the minutes and seconds from those used in time determination. The conventional symbols are ° for degrees, ′ for minutes of arc (but *not* minutes of time) and ″ for seconds of arc (again, this should not be used for seconds of time). Thus 10 degrees, 20 minutes of arc and 10 seconds of arc are more conveniently written 10° 20′ 10″. The reader, as well as the author, will doubtless find the subsequent use of this convention simplifies matters.

emphasis again shifted back to the stellar universe. But now it was concerned with the nature of the objects that could be seen – many of them strange and diffuse and clearly not stars, whatever else they might be. The old time-honoured question of position determination was not neglected, although its aims included determination of the newly discovered stellar motions and the solution of the long-standing problem of finding stellar distances. At last optical refinements brought success, and it was possible to measure accurately angles as small as one third of a second of arc – the equivalent of being able to detect a pinhead at about 700 yards.

The first successful determinations of stellar distance were made in 1838, and within the next few decades the development of yet another technique was to increase the accuracy of astronomical measurement and thus the distances that could be directly found; it was also to allow the astronomer to examine his results at leisure and with detailed scrutiny, and let him probe further into space and examine what he saw there with much greater ease. This was the coming of photography – a new invention aimed originally to assist the artist and amuse the public. Its applications have affected every branch of science, yet none more profoundly than astronomy. From the point of view of accuracy the photographic plate permits the astronomer to measure his recorded telescopic image with a microscope, and so attain a precision greater by a factor of a hundred than was possible in 1838. But on the purely observational side, the improvement has been equally dramatic. There is a threshold of sensitivity below which the human eye cannot operate and, in consequence, there is a limit to how dim an object can be if it is to be detected, even with a telescope. With the world's largest optical telescope at the time of writing – the 200-inch aperture reflecting instrument on Mt Palomar in California – the dimmest star that can be detected visually is one with a brightness or magnitude* of about 18. This is an increase in sensitivity of some 60,000 times that possible with the eye alone and is due to the much vaster area of the 200-inch mirror compared with the aperture of the pupil of the human eye, which is no more than $\frac{1}{4}$ inch. Yet the use of photography brings greater sensitivity still, since image building on a photographic emulsion is a cumulative process – the longer a plate is exposed the dimmer the objects it can record – whereas the eye has no such power. With a special plate designed

* The reader may care to be reminded that astronomical magnitudes have nothing to do with size but are solely a measure of brightness. Numerically they are a scale of dimness, thus a star of magnitude four is dimmer than a star of magnitude three, which is itself dimmer than a star of magnitude two – and so on. On a scale of this kind, negative magnitudes have to be used for very bright objects.

for astronomical work, the 200-inch will show objects with a magnitude of no more than 23·9, an increase in sensitivity over visual observations through the same instrument by a factor of 100. It is no exaggeration, then, to claim that it is photography that, with its additional power to record for later repeated examination, has made it possible to amass observations which have not only taken the astronomer into regions of space hitherto inaccessible, but by doing so have helped substantially in the formulation of our present ideas of the universe as a whole. Objects can be photographed that were previously invisible and they confirm and extend a belief based on visual observations – that the universe extends as far as we can see; their details can be examined and analysed also in a way that is impossible with the eye alone, since the photographic emulsion can be made more sensitive to a wider range of colours or wavelengths than the eye. In one sense photography enters the realms of what we may define as 'invisible astronomy' for it shows us objects that cannot be seen by the eye and the telescope alone.

A further extension of our appreciation of the universe occurred when, in the mid nineteenth century, studies began with a new instrument – the spectroscope – designed to disperse sunlight into a spectrum and so display its separate colours. This coloured spectrum, known for ages in the form of the rainbow, was imperfectly understood until the time of Newton, yet even then it was not for a century and half that its examination was to begin in detail. Theoretical work in the laboratory and the extension of spectroscopic methods to the stars and, later, to other celestial objects, made it clear that it was possible at last to make some analysis of their chemical constituents. But the spectroscope was soon found to provide more than a means of chemical analysis. It could give a value for the temperature of the surface gases of stars, a clue to other physical conditions such as the pressures and densities to be found there and, above all, a measure of the line of sight motions of all celestial objects. This was to prove of vital importance, for although tangential motions (movements across the sky) can be observed by comparing observations made on different dates, all objects except the planets are too far away to show any diminution of size indicative of a motion away from us in the line of sight. One might conjecture that such motions were likely, but without observational evidence it could be no more than a guess. With the spectroscope, however, the situation was completely altered. The position of the lines that cross the spectrum and are characteristic of different chemical elements could be measured and were found displaced from their positions as observed in the laboratory. This, Christian Doppler suggested and Hippolyte Fizeau showed, was due to a motion towards or away from the observer – towards if the shift was in the direction of the violet end and away if it moved nearer the red end of the spectrum. Moreover, the degree of shift gave a measure of the velocity along the line of sight. This was of profound significance, since not only did it make possible the determination of the true way the stars moved within our own Galaxy – but also the way in which other galaxies were moving. The result of this was the revolutionary hypothesis of the late 1920s that the whole universe is expanding outwards, a concept which we still accept, although with some reservations, as we shall see in Chapter 10.

The reason for this brief sketch of some of the main turning-points in observational techniques is to emphasize the fact that what we understand of the universe – or rather the raw material on which we base our theories and on which we can exercise our imagination – depends upon what we can observe. This may seem so obvious a truism that it hardly needs saying, let alone emphasizing, yet surprisingly enough there are those who consider observation at best a guide and, on occasions, an unnecessary encumbrance. Their argument seems to run as follows: mathematics is a precise form of logic and by sitting down and thinking about the physical universe, and using a minimum of basic facts, one can construct a logical mathematical system of equations which will give us a universe as it essentially is. We may need to interpret our results and use observation as a guide for fitting figures to our equations, but this is the limit of observation's role. Essentially the universe must be a fabrication of logical thinking, and if a statement is logical and reasonable, it must be accepted. This rationalist approach can be a powerful method, but it falls – or is likely to fall – into the danger of asserting that something is so if no evidence against it can be produced. For instance if we assert that matter is being continuously created throughout space at a rate which will satisfy theoretical considerations but is too small to be directly observable, then no direct scientific results can be obtained to refute this idea. Or again, if there are parts of the universe that cannot be observed, then since it is impossible to adduce scientific evidence to the contrary, we can claim that such parts must be similar to those we can record, and we can claim this without any fear of contradiction. Such an approach is not new: it was used with great ingenuity by many of the natural philosophers of ancient Greece and it is a powerful method, especially with the recent development of new mathematical techniques. But to prevent the construction of 'impossible' universes, the rational theorist must temper his views by having recourse to observation. Even so, he still has plenty of room in which to manoeuvre, since many observations require interpretation to determine their physical implications: theory, too, may lead astronomers to make only some observations and avoid others. The one basic error is to imagine that a unique result can be obtained in this way and that observations can be legitimately interpreted in whatever way appears most satisfactory for the theory in hand.

On the other hand, we must take care not to accept all observations at their face value. The red-shift of spectral lines can be interpreted as a recession away from us. Observations of the rotating mass of the Sun show a red shift of that edge or limb which is moving away and a violet shift of the limb that is approaching us, and in this instance there can be no doubt about the cause of the shifts. But when we come to apply this to the most distant parts of space, where forces may be operating to a degree unknown or unfamiliar in objects close to us, it is important to appreciate that the same interpretation *may* not hold. We must be prepared to consider that there can be more than one way of interpreting an observation and that some empirical fact that has previously seemed to be comparatively unimportant, may later turn out to be of considerable significance. Radio observations made by Karl Jansky in the United States in 1931 showed that there appeared to be a radio transmitting source in the direction of the constellation of Sagittarius, but his research was primarily directed to studying the crackling or 'interference' that interfered with short-wave radio reception and little notice was taken of his discovery. Yet later it was to prove the foundation stone on which the whole of the new subject of radio astronomy has been built.

The truth of the matter is that we must try to strike a balance between theory and observation, being neither too rationalistic in outlook on the one hand, nor too exclusively empirical on the other. It has been said that the difference between the two viewpoints can best be summed up in the quip that since he can always obtain a great number of solutions of his equations, a rationalist will require an infinite number of points before he can draw an unequivocal graph of the relationship between two quantities, whereas the empiricist, with the experience of all his experimental knowledge to fall back on, will only require one!

But if we are to strike a balance between observation and theory, and particularly if we are to interpret novel observations in any way correctly, we must realize that one of the fruits of any theory worthy of the name is the fact that its consequences should lead us to expect quite specific results from observation. Indeed a good theory should go so far as to suggest that if we make certain observations, we shall obtain particular kinds of result. Newton's theory of gravitation did this, the most famous 'prediction' it led to probably being that computed by Edmond Halley, who in 1705 published details of the return of a bright comet in 1758, last seen in 1682. Again, in 1916 computations using Einstein's general theory of relativity led to the expectation that rays of starlight would be deflected when passing close to the Sun by an amount twice as great as that to be expected on Newtonian theory, and in 1919 observations made of the dark sky during a total solar eclipse showed that this was precisely so. A theory requires observational or empirical confirmation, and observations need a theory to turn them from a mass of unrelated data into a correlated body of evidence. Theory and observation must complement one another, even though at different times it may be either observation or theory that advances the more rapidly.

While on the subject of theory and observation, we should spend a moment on the question of the laws of nature and their applicability throughout the universe. After all astronomy, above every other branch of natural science, stretches out to the known limits of the physical world, and it is necessary to question the universality of laws which we have discovered from terrestrial observation alone, or from observations made comparatively close by as far as astronomical distances are concerned. How correct is our assumption that what applies on Earth applies equally well at distances so vast that light has taken thousands of millions of years to reach us? How binding is a scientific law?

First of all, it will be as well to realize that a scientific law is not a law in the sense that it is a divine right that cannot be gainsaid – indeed it might save considerable confusion if the word 'generality' was used instead of 'law'. No scientific law is ordained by Nature or by anything or anyone else; it is nothing more nor less than a statement of generality, induced from a host of separate observations, and a generality that the scientific community has accepted after examining and correlating observations, theory and experiments. It expresses the behaviour of things and owes something to theory as well as to observation: in essence it is a synthesis of both. And no scientific law is necessarily immutable. It may hold sway for a very long time, just as the law of uniform planetary motion in a circle held sway for more than 2,000 years after the Greeks had proposed it, but this is no guarantee that it is eternally true. Ever since the early seventeenth century when Johannes Kepler demonstrated that planetary motions are elliptical, a different 'law' of planetary motion has been accepted: one that better fits the more precise observations made after Greek times, and agrees satisfactorily with a new and more widely embracing theory of planetary motion. Yet the concept of uniform circular motion had become so ingrained in peoples' minds that its overthrow seemed a revolt against reason. A similar attitude occurred after the heliocentric theory that placed the Sun in the centre of the universe supplanted the old time-honoured view that the Earth lay at the centre of all things – a revolution in thinking occurred and laws of physics that seemed permanently established were in due course replaced by new laws. Objects no longer fell to the ground because they were seeking their natural place in the universe, but because of a law of universal gravitation. Water spread around the Earth not because of its divinely ordained place in the scheme of nature but, again, because of gravitation. The new law was applicable, too, to the whole solar system and probably beyond into the realm of the stars: it was more comprehensive than any that had gone before and was accepted in the scientific world, even though no one could explain the nature of gravitation itself – that it was a force operating between bodies had to be sufficient. And what occurred in these cases, occurred in other fields of natural knowledge, indeed the whole history of scientific thought shows that in no case must we consider a law as unchangeable, as permanently part of the scientific picture. Observations or theories may alter or replace it so that something more widely embracing, more in accord with new knowledge and more satisfying, may supplant it. We shall still have a generalization, but one that has advantages over that previously accepted. What we must not fail to accept is that even the most recent law may be modified.

The question of the nature of scientific laws is not purely an academic exercise, for in considering the universe as a whole it has particular relevance. For instance, we must consider the question of applicability, since laws of physics, which play a large part in modern astronomy, are developed for terrestrial conditions. The law connecting the pressure and volume of a gas in its simplest form – the so-called Boyle's law – was a law derived from experimental work in terrestrial laboratories. To begin with this was its sole foundation and it was no more than what is sometimes termed a phenomenological law – a law derived from observing phenomena. This is often the way with a discovery and only at some later date does theory catch up with experiment and embrace the phenomenological laws into laws of theoretical science. In the case of Boyle's law, it became incorporated into the molecular theory of gases when this arrived some two centuries later. But how far can Boyle's law be applied? It is relevant to phenomena on Earth, but is it still valid when we come to consider the behaviour of gases on the Sun? And if it is applicable there, can it be extended to other stars, or to other galaxies? Where, if anywhere, are its limits of application? Presumably, now that molecular theory accepts the law as one of its consequences, it can justifiably be applied to any place where molecules are to be found, and if this is so, then it may be legitimately used wherever we use physical science, be it on Earth or out in the depths of space. Clearly, we cannot *prove* that this is a correct assumption, but it appears logical enough.

Gravitation can be cited as another example. When Newton formulated his law of gravitation, he did so as a synthesis of experiment, observation and theory covering terrestrial phenomena and planetary motion, but the question arose, as it was bound to, whether the law was universal enough to apply to the stars as well. Newton himself conjectured that it did, and his colleague Halley took the same view, for they both believed that this was a legitimate extension of the theory. Yet it was not until a century later, with William Herschel's discovery of double stars in which the two members orbited round one another in accordance with Newton's theory, that some independent evidence for this extension was forthcoming. After that no one seemed inclined to query the extension of the law of gravitation as far out in space as may be desired. But as with Boyle's law, we have no *proof* that it is universal: it only seems a more likely hypothesis than to apply arbitrary limits.

When in the late nineteenth century the study of spectroscopy began, it became possible to analyse the light emitted by glowing bodies and determine their chemical composition. This technique was soon applied to the stars, notably by William Huggins and Angelo Secchi, and it seems to have been tacitly assumed that this was perfectly legitimate. In support it may be noted that interest in the spectrum and the lines across it had been stimulated by Fraunhofer's detailed examination of the solar spectrum in 1815, and so the extension of later laboratory work to observations of the stellar system would appear proper enough but, as in the other cases quoted, there was no proof that this was correct; it was merely considered – and not without good sense – that to refuse to apply the new knowledge in this way would have been perverse. Only in 1895 when William Ramsay discovered and isolated helium in the laboratory, and identified it with an element known by its spectrum lines to exist in the Sun, was something tangible available to support the new technique.

The extension of terrestrial physics to celestial situations is now an accepted part of astronomical research, and our astronomical knowledge would be the poorer if this were not so. The most important thing about it all is that we should realize what we are doing, so that when any crisis arises, when anything occurs that seems to be an exception to our laws of nature, we once again go over the ground, examining each stage in our arguments, each extension of terrestrial physics, so that we may discover what factors are involved. Only thus shall we be able to see what is happening and perhaps discover new and important factors that may be in operation; factors that are insignificant in their terrestrial effects over the small distances involved on Earth, and which only become large enough to observe over, say, the vast spaces between one galaxy and another. Indeed, the hypothesis of 'cosmical repulsion' to account for the continued motion of galaxies away from one another is just such a factor.

The general applicability of our local laws of nature to the universe at large is sometimes known as the *uniformity of nature*. As already indicated, there appears no way of proving it although, when astronauts visit the Moon and the nearer planets, it will perhaps be legitimate to consider that it has received some independent confirmation. Yet it is so vital a principle, so well in accord with the general scientific outlook and the widespread belief that we live in a rational and not a capricious universe, that its acceptance should cause no qualms. On some occasions it has been referred to as the scientists' 'article of faith' but, if this is what it is, then it certainly has more practical results to support it (not prove it, of course) than articles of faith usually accepted as such. And hand in hand with the uniformity of nature goes a second postulate – at least in astronomy – and that is the *cosmological principle*. This states that the universe appears the same to any observer, on whatever galaxy he may be situated. It is in one sense an extension of the principle of uniformity and, in another, an extension of the view that came after the Earth, and then the Sun, had been dethroned from their privileged positions in the centre of the universe. We have no reason to suppose that the Galaxy of which the Sun is a member is in any preferential position and, once we accept this, we cannot argue that there is anything special about our scientific observations. Those obtained by observers on a planet orbiting a star in some other galaxy should be expected to be similar: hence the enunciation of the cosmological principle. The only restriction on it is that any other galaxy shall not be moving at a much greater velocity with respect to others than is our own, for then the universe observed from it would appear different, or so the general theory of relativity states. Yet even though the appearance would be different in such a case, the laws of nature should not alter since those which such an observer would formulate would be the same as ours provided relativity corrections are taken into account.

An attempt to extend the cosmological principle has been made in recent years – from 1948 to be precise – by Professors Hoyle and Bondi, then colleagues at Cambridge University. They have stated that the universe not only appears the same from whatever galaxy observations are made, but also at whatever time the observations may be taken. In other words they claim that the universe is eternally the same. They do not preclude change; indeed they agree that galaxies are born, develop and die – as all astronomers believe – but that overall the universe is changeless, new galaxies replacing those that cease to exist or cease to be visible. This extension they call the *perfect* cosmological principle. Their view has aroused considerable opposition, and their assumption of a static cosmological principle has even been termed the perfect cosmological presumption, allied as it has been with a theory of the universe that is far from generally accepted. Yet whatever may be the fate of the theory – and as will become evident in Chapter 10, it seems possibly doomed to rejection – there is an interesting and worthwhile point behind their proposal of a 'perfect' cosmological principle. It is simply the fact that if the universe changes markedly with time if, say, the galaxies were all closely packed together at some remote epoch in the past (a possible corollary of the idea that the universe is expanding outwards) – then there is nothing to lead us to suppose that laws of nature which are valid in conditions as we now find them, would have been valid under such different conditions.

There is no doubt that the perfect cosmological principle is a powerful implement with which to probe the essential nature of the universe, but equally there is no proof that the laws of nature would be changed in essence in a contracted universe. It may be safer to assume that they would, but that would appear to be all we can say, and at the present time it seems to offer some very severe restrictions on the kind of theory we are free to make about the universe. Whether it is valid or not, we certainly have insufficient evidence to hand to allow us to decide, and to many astronomers it appears an unlikely possibility. To be fair, however, the decision must remain in abeyance, at least for the present.

Before leaving the question of general assumptions and basic premises about the universe and the scientific laws we apply to it, there is one further matter that must be mentioned, and that is the limitations imposed by the velocity of light. The velocity of light was first measured in 1675 by Ole Rømer who obtained his value from observations of hitherto unexplained delays in the interval between eclipses of the satellites of Jupiter as they orbited the planet. Later experiments have provided improved values but Rømer's observations led to the realization that the velocity of light is finite, and thus the enunciation of a basic and important astronomical principle, since all our empirical evidence is based on the receipt of light or of radiation at other wavelengths which still possess the same velocity. The limitation that this imposes arises from the expansion of the universe, since it seems well enough established that galaxies travel away from us with velocities that increase with distance, and therefore there should come a point where the velocity of a galaxy reaches the velocity of light. At this point radiation emitted from such a galaxy can never reach us and, in consequence, it must remain for ever unobservable. If this hypothesis is true then there would indeed be a limit to the observable universe, the limit being a sphere within which the galaxies are all travelling at velocities less than the velocity of light. But in fact this hypothesis occurs in one mathematical scheme of the universe – the 'model' proposed by the late Abbé Lemaître, in which the universe began with a giant explosion and expansion continues due to cosmical repulsion – while there are a few others, that have different limits or horizons. Yet if we accept (as almost every astronomer does) the theory of special relativity, we not only have a horizon that depends on our model, but also need to face the fact that the relationship of red-shift and distance is not one of simple proportionality. The exact relationship we adopt depends on what mathematical scheme of the universe we accept, and this ambiguity arises both because observational evidence is not yet sufficient for us to choose one particular scheme, and because the distances of objects very far away in space cannot be determined with sufficient accuracy. What can be said, however, is that at great distances the red-shift takes us to velocities that approach the velocity of light but never quite reach it. Such limitations as there are in the observation of distant objects appear more likely to come from the limits of the sensitivity of our equipment rather than from any inherent property of the velocity of light, unless we accept the view that the universe began a finite time ago. In that case the observable limit is that distance from which light would take longer to reach us than the age of the universe, and in Lemaître's universe there is the additional factor that at a certain still nearer distance, cosmical expansion causes the velocity of galaxies to become greater than the velocity of light. At all events, wherever the horizon lies – if it exists at all it is from the light and other radiation we receive that our picture of the universe is built, and the mathematical scheme we eventually accept will be chosen.

To sum up, then, while in theory a model of the universe may be constructed from a few basic postulates, it is generally agreed that any model should be checked by observation and, in general, that observational results are really the foundations on which any theoretical scheme should be built. And it is from observational results coupled with theory that our scientific laws are formulated. These laws may change as our theoretical knowledge, our observational techniques, or both, develop. We assume that these laws apply throughout the universe: an act of faith which cannot be proved but without which we cannot pursue astronomy or any other branch of extra-terrestrial physics. Lastly, all our information about the universe comes to us either in the form of light or other radiations, all of which travel at the same velocity through space. These are the bed-rock of our subject; they require interpretation but they are the basis of all our ideas. As our techniques for receiving them increase, so our conception of the universe becomes broader and more complex: the small Earth-centred world of Man gives place to a wider, more impersonal universe.

Recently new dimensions in our understanding of the universe have opened up owing to observations of radiation that has hitherto been inaccessible to our instruments, and so far the results and their interpretation have brought some astonishing re-assessments of our knowledge. Added to the normal visual radiations, these invisible wavelengths are proving of such importance that it is hard to over-estimate their value. What they are, how they are observed, and how they are generated will be discussed in subsequent chapters. In the last chapter, an attempt will be made to assess their present significance.

CHAPTER 2

The Electromagnetic Spectrum

Light is something intangible yet ubiquitous, and it is little wonder that its nature eluded natural philosophers for millennia. It was only after the mid-seventeenth century that physical theories about it were formulated, although its behaviour in terms of reflection and refraction had been understood for some years before. Two theories claimed the attention of physicists – one, favoured by Newton, that considered light primarily as a movement of small particles or 'corpuscles', and another, of which Christiaan Huygens was the chief architect, which conceived of light as a wave disturbance. Today we have progressed somewhat further, although even now we have to support a kind of coalition hypothesis, since in some experiments light behaves as if it were composed of particles, and in others it leaves no doubt in the mind of the experimenter that it is indeed a wave disturbance. Light is therefore described as being composed of photons, each of which can best be thought of as a 'packet' of waves. This appears, on the face of it, to be an unsatisfactory idea, but it does correlate the empirical evidence and can be expressed in quite elegant mathematical terms.

However, theories apart, it was that great observational astronomer William Herschel who, in March 1800, reported to the Royal Society on 'radiant heat' and thereby drew attention to the extension of the coloured spectrum into a range quite invisible to the eye. This and subsequent events must now be considered briefly, even though they may seem to be a digression, because in this way the essential ideas behind our modern views can best be understood.

Herschel had been led to his investigation after experiments with dark glass of different colours as a means of reducing heat in solar observing, and he had noticed that some allowed more heat through than others. In his investigation he passed a narrow beam of sunlight through a prism and allowed the dispersed beam to fall on a broad slit. From thence isolated colours could be passed on to fall on to the bulbs of a number of thermometers, and as a result he found that while red light appeared to have some heating effect, the change of temperature was greater when the thermometer bulbs were irradiated by what he assumed was radiation coming from below the red end of the spectrum. Later experiments, reported soon after, allowed him to assert that this invisible radiation could be refracted and reflected in a similar way to visible light. Yet Herschel soon abandoned the hypothesis that heat radiation was an extension of the visible spectrum, and only later was this view generally accepted. Meanwhile, in 1801, Johann Ritter had found that radiation appeared to exist beyond the violet end of the spectrum since, although this exerted no heating effects, it could cause chemical changes to occur such as the blackening of a light-sensitive material like silver chloride. However, the nature of these invisible radiations was unknown.

It was also at the beginning of the nineteenth century that the wave theory of light was resuscitated, first by Thomas Young and then by Augustin Fresnel. Huygens had thought of light waves as pushing their way through space, much as sound waves do through air, but to account for all the empirical results, Young and Fresnel had to abandon this explanation and conceive of light waves as an up-and-down or transverse disturbance, analogous to waves in the sea. This was in 1816, yet it was still to be some years before a comprehensive theory of radiation could be established. Empirically the next major step was probably the demonstration by Michael Faraday in 1834 that light could be affected by a magnetic field. Faraday, who had by this time investigated a host of magnetic and electrical effects, had put forward the concept of a magnetic 'field' to rationalize the connections he had discovered between magnetism and the flow of electric currents. In another investigation he had developed a very dense kind of optical glass, and had begun to seek relationships between electricity and other phenomena. His friend John Herschel (son of the astronomer William Herschel), who had been studying mineral crystals and their ability to affect light, had conjectured that similar optical effects should be associated with electrical and magnetic phenomena, and this stimulated Faraday's experimental genius to try to observe them.

The effect John Herschel and Faraday expected was a rotation of the plane of polarization of a beam of light. This was essentially linked in their minds with the new wave theory of light. If we consider a beam of light in which the vibrations are transverse, then initially the plane of the vibrations may lie in any direction (figure 1) but by passing light through a polarizing substance such as the transparent mineral Iceland spar, light vibrating in one plane only is transmitted. Faraday's experiment consisted substantially of two double prisms of polarizing material with a piece of very dense glass in between, and a strong electromagnet

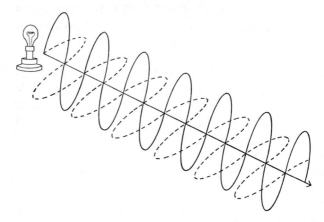

Figure 1

Unpolarized light. The transverse electromagnetic vibrations of which light is composed can lie in any plane when unpolarized. Here, to avoid confusion, the vibrations are shown in two planes only, by the full and dotted lines.

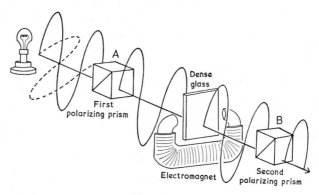

Figure 2(a)

The essential parts of Faraday's experiment to investigate the rotation of the plane of polarization by a piece of dense glass and an electromagnet. Unpolarized light enters from a source (for convenience shown here as a filament lamp although this was invented after Faraday's time); it passes through the first polarizing prism (A), thence through a sheet of dense glass and to a second polarizing prism (B), which 'analyses' the beam and, since it is polarized in a vertical plane, passes it through. The electromagnet surrounding the dense glass has not been switched on.

I. The 200-inch optical telescope at Mt. Palomar. A view looking down the tube
of the telescope. The 200-inch (5 metre) diameter mirror can be seen at the far
end of the tube. The prime focus cage, to which the light from the 200-inch
mirror is brought to its primary focus, is at the top of the tube. In the cage, an
observer is adjusting a camera.

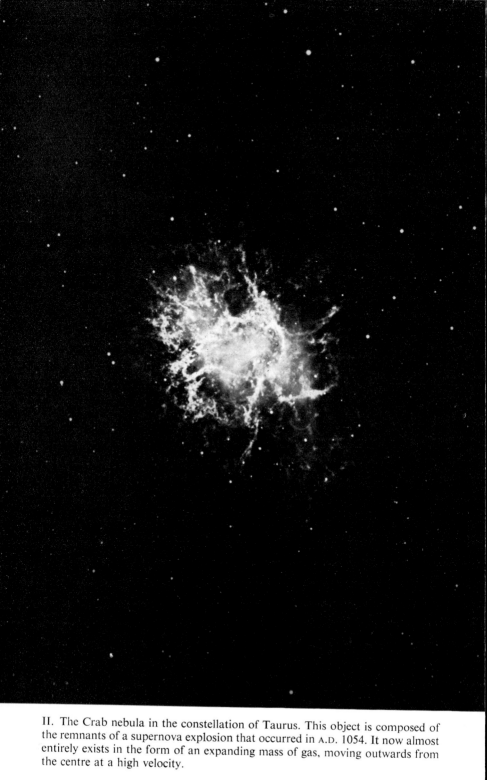

II. The Crab nebula in the constellation of Taurus. This object is composed of the remnants of a supernova explosion that occurred in A.D. 1054. It now almost entirely exists in the form of an expanding mass of gas, moving outwards from the centre at a high velocity.

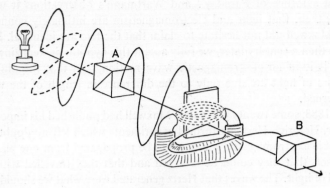

Figure 2 (b)

The electromagnet is now switched on. The magnetic field between its poles (shown by dotted lines) now affects the dense glass and causes the latter to rotate the plane of polarization (here shown rotated through a right angle), of the polarized light passing through it. The new plane of polarization is unacceptable to the polarizing prism (B) and no light emerges.

(figure 2a). With no magnetic field the polarized light passed from A to B and out to the observer, but when the electromagnet was switched on and its field arranged to lie parallel with the plane of polarization of A and B (figure 2b), then the plane of polarization of the light was rotated as it passed through the glass; in consequence the polarizing material B would not permit the passage of the light beam and the observer saw no light. This was an important result which was confirmed the next year by the Swiss physicist E. E. Wartmann, who achieved similar results with William Herschel's heat rays (infra-red rays), and then by others.

This empirical evidence now needed welding into a coherent theory, and credit for this really lies with James Clerk Maxwell who, in 1864, published a set of mathematical equations that expressed the relationships between electric currents and magnetism which Faraday had discovered in his brilliant experimental work at the Royal Institution. Two very important facts emerged from these equations. First, the equations themselves were similar to those which express the behaviour of light waves, and so could be taken to describe the behaviour of waves, in this case waves with electric and magnetic effects – electromagnetic waves. Second, when the velocity of these electromagnetic waves was calculated (from the values of the electrical quantities involved in the experiments), it was found to be the same as that for light. With the

c

added evidence of Faraday's and Wartmann's observations it would seem, then, that light and electromagnetism are intimately connected and Maxwell did not hesitate to claim that the two are identical. Now, more than a century later, we fully accept this hypothesis, realizing that light is itself an electromagnetic wave motion, and that the different colours of light are due only to the difference in length of the waves concerned.

In 1888, some twenty years after Maxwell had published his important paper, Heinrich Hertz performed experiments which left no doubt that invisible electromagnetic waves could be propagated from one place to another, that they could be reflected and that they travelled with the speed of light. The waves that Hertz generated were what we should now call short radio waves and they differed from light waves or infra-red waves only by reason of their length. In the years that followed studies were extended and the theoretical construct of the electromagnetic spectrum was formulated.

The coloured spectrum of light is only a part, and a very small part, of the whole electromagnetic spectrum. The lengths of light waves are small and are usually measured in Ångströms (written Å)* or in milli-microns (written mμ); one Å is one hundred millionth of a centimetre (10^{-8})† and a millimicron one millionth of a millimetre (10^{-6}mm). Deep red waves have lengths of 7500 Å (or 750mμ) and violet waves of 3900 Å (or 390mμ) – hence red waves are longer than violet, and the general rule is that wavelengths decrease from red down the spectrum to violet. Below the violet, still shorter wavelengths are to be found, first the ultraviolet rays discovered by Ritter, which range from 3900 Å down to 140 Å, then X-rays which can be as short as 1 Å. Even so the range of short wavelengths is not exhausted and gamma rays (written γ-rays) with lengths less than 1 Å are found, with a lower limit somewhere about 10^{-5} Å.

At the longer wavelength end of the spectrum, beyond the red, come the infra-red rays or radiant heat rays of William Herschel, with wavelengths ranging from about 7500 Å to something of the order of 3×10^5 Å or, in more readily familiar terms up to wavelength of $\frac{1}{10}$ mm ($\frac{1}{250}$ of an inch). Beyond this there come the radio wavelengths that stretch

---

* Named after the Swedish physicist Anders Ångström (1814-74) who carried out much pioneer work with the spectroscope.

† For readers unfamiliar with this notation, it should be explained that it indicates a division by the power given. For instance $10^2 = 10 \times 10 = 100$ and $10^3 = 10 \times 10 \times 10 = 1,000$, but $10^{-2} = \dfrac{1}{10 \times 10} = \dfrac{1}{100}$, $10^{-3} = \dfrac{1}{10 \times 10 \times 10} = \dfrac{1}{1000}$, and so on. By convention $10^0 = 1$.

from the fractional millimetre range up to the immense size of 10,000 kilometres – that is wavelengths of some 60,000 miles. The known range of the electromagnetic spectrum is, therefore, vast, with the longest (radio waves) some 10^{22} (or ten thousand billion billion)* times greater than the shortest γ-rays. Of this immense spread of wavelengths, light covers only some three thousand billion billionths and, in consequence, Man's natural sensitivity to the electromagnetic spectrum which he has used for examining the universe is small indeed compared with the immense range that he might be expected to observe – that is, of course, assuming that objects in the universe do, in fact, radiate at all electromagnetic wavelengths. Yet even if the radiation covers only some of the other non-visual wavelengths, it is clearly important to observe in these wavelengths and obtain the additional information that they may be expected to provide.

The reader cannot fail to have noticed that the figures for the various wavelength ranges are somewhat vague, except for the range of visual wavelengths. This is because the names are somewhat arbitrary since they were originally coined partly according to where they lay with respect to the visual spectrum (ultra-violet and infra-red), or the means used for detecting them (X-rays, radio waves), or the name originally given them in a quite different connection (γ-rays). The fact that these ranges overlap – some very short ultra-violet wavelengths being in other contexts referred to as very soft X-rays, for example – means that precise nomenclature is not possible. Ambiguity can only be avoided by referring to the wavelength or, as the physicist and the radio astronomer sometimes finds more satisfactory, by the frequency of the radiation.

The concept of frequency is simple enough, and is a measure of the number of undulations or waves passing the observer during each second of time.† Clearly, the shorter the wavelength the closer together the crests of the waves are to be found (figure 3) and the more that will pass each second, since all electromagnetic radiation travels at the same velocity. In other words, the higher the frequency the shorter the wavelength. In radio physics frequency is measured in units known as Hertz (Hz), 1 Hz being a frequency where one wave crest passes the observer per second, and the radio range of frequencies runs from some 3 million million Hertz (or, as more usually expressed 3 million MHz [Mega

---

* The billion used here is that adopted in the United States – namely one thousand million.

† The mathematical equation linking frequency (ν), the wavelength (λ) and the velocity of light c is simplicity itself, since $\lambda\nu = c$. The value of c is 186,300 miles per second or 3×10^{10}cm per second.

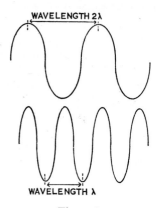

Figure 3

Wavelength. Of the two waves shown, the upper has twice the wavelength (2λ) of the lower (λ). The frequency of the upper wave is half that of the lower since, as both move with the same velocity, the number of crests (and troughs) passing a receiver from the wavelength λ is double that from the wavelength 2λ.

Hertz]) to 30 Hz. In some aspects of invisible astronomy it will prove more convenient to use frequencies rather than wavelengths.

In discussing the range of the electromagnetic spectrum, no mention has so far been made of cosmic rays, the existence of which was first discovered in 1909 by the physicist Göckl. What Göckl and the others have called cosmic rays are not, however, rays at all, but high speed atomic particles. Confusion of nomenclature arose because the particles were not noticed and only the short wavelength radiation to which they gave rise was observed. Cosmic particles will be examined in Chapter 7, but the radiation to which they give rise has already been considered since it is included in the γ-ray range already mentioned.

But if one effect of cosmic rays can be included in the vast range of the electromagnetic spectrum, it is convenient at this stage to limit ourselves to an examination of the visual spectrum: a very restricted part, certainly, but one in which much has been discovered. Indeed, the visual spectrum has provided the general picture of the universe with which the astronomer is familiar. Since it is visual, the details and concepts that have grown from it are more readily appreciated and can form a useful foundation for the analysis of the research being carried out in the invisible wavelengths.

The visual range of the electromagnetic spectrum, running from red to violet, gives us the range of colours that are observed in the heavens and, in particular, the range of colours of stars, which vary enormously. Yet by themselves these colours provide little information, and it is only when the visual radiation is dispersed with a spectroscope that full use can be made of it. The spectroscope is essentially a device for sorting out the various wavelengths and in order to achieve this uses a prism or a grating. A prism is able to separate the visual wavelengths, because although all electromagnetic radiation travels at the same speed this is only strictly true if it is travelling in a vacuum. When it enters a transparent substance such as glass, the velocity is altered and, what is more, the change of velocity is different at different wavelengths, the shorter

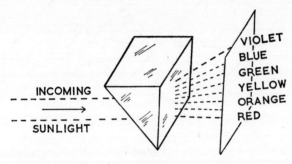

Figure 4

The action of a glass prism in dispersing white light into its component colours (wavelengths).

wavelengths being reduced in velocity more than the longer ones. The result of reduction of velocity is to turn the waves of the incoming visual radiation through an angle and thus alter their direction: this alteration is more for short wavelengths than for long; in consequence the blue and violet rays lie in a different direction from the orange and red, and indeed the whole conglomeration of wavelengths is spread out into a coloured spectrum, as Newton discovered (figure 4).

A grating disperses light into its separate wavelengths by making use of the phenomenon of diffraction. Diffraction, which can only be explained satisfactorily on the wave theory, is the technical name for the bending of light round objects, and can best be described by an analogy. Suppose waves are coming in from the sea and meet a breakwater, then they will curve round and spread out into the space either side of the breakwater. A similar kind of effect may be observed if an obstruction containing a small opening is placed in a tank of water (figure 5); waves set up on one side of the tank (A) will be stopped by the obstruction except at the aperture (B). Here they can pass through and, after doing so, will spread out into the calm water (C). This phenomenon occurs in light and can be observed experimentally; it was investigated some three centuries ago by the Italian Francesco Grimaldi, who found that the passage of sunlight through a narrow aperture and past an obstruction gave coloured fringes to the shadow, and that passing light through two small apertures also provided a disk of light

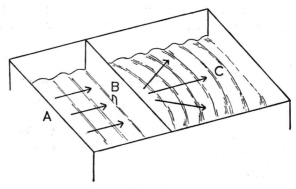

Figure 5

The phenomenon of diffraction demonstrated in a tank of water. Waves set up in the left-hand compartment (A), pass through a small aperture (B) in a dividing wall to the right-hand compartment (C). The waves from the left are diffracted by passing through B and spread outwards. The amount by which the waves spread out depends upon their wavelength.

on a screen that was larger than that to be expected if light were supposed to be composed of particles travelling in perfectly straight lines. Under ordinary conditions diffraction goes unnoticed, since the wavelength of light is so small and the diffracted light is dim, and it is only when it is passed through very small apertures that the phenomenon is observable.

Since different colours have different wavelengths, the way in which each wavelength behaves on passing through a small aperture will differ, and if a beam of sunlight, for instance, is passed through, the colours will be separated, since the different wavelengths will spread out by different amounts. This was one of the effects observed by Grimaldi, and it can be utilized for obtaining a spectrum. It would, however, be inconvenient to use one small aperture since the amount of light available for examination would be seriously reduced, and the spectroscopist makes use of a grating. This consists of a piece of glass with a great number of straight lines ruled across it. The lines are very close together – usually there are some 15,000 to the inch – and thus the apertures they form are small compared even with the wavelength of light. When light passes through the grating each colour spreads out by a different amount. At specific directions, the light of one wavelength from one aperture will reinforce the light of the same wavelength from other apertures and the observer will see a particular colour. He will see this colour in considerable purity because the rays of light of other wavelengths observed in that particular direction will have spread out from the apertures in such a way as to cancel one another out. Or, to put it another way, as the light spreads out into its different wavelengths from the separate apertures, these will either reinforce or destroy themselves, and which they do will depend upon wavelength and direction. There will be one direction for each wavelength where reinforcement occurs and so the observer, whose field of view embraces many of these directions, will see many colours – a complete spectrum in fact.

The phenomenon of interference, which is of considerable use in invisible wavelength observations, especially in radio astronomy, as will become evident in Chapter 6, is simple enough in principle. For consider radiation of a particular wavelength (figure 6a): when two beams of this radiation are travelling in step with one another – in phase, as it technically is called – they will reinforce one another (figure 6b), but when they are exactly out of phase, the undulations of one will cancel the undulations of the other (figure 6c) and no radiation is observed. In a diffraction grating, the phenomenon provides an efficient method of dispersing light into a spectrum, but in a different design -- the stellar interferometer – it can be used to measure the diameters of large bright red stars. In such an instrument, diffraction also plays its part in that it causes star images to appear as disks, each surrounded by a number of dimmer concentric rings. Such disks are spurious in the sense that they are purely optical effects and bear no relationship to the real diameters of the stars. However by using an interferometer consisting of two small apertures placed at the front of the telescope, each star gives two separate images and thus two separate spurious disks.* These disks overlap and, in consequence, the light waves composing them interfere, so that the observer sees a spurious disk crossed by dark

* A new type of stellar interferometer has recently been designed, using two reflecting telescopes on rails, whose separation can be varied at will. The light is observed through special interference filters using electronic techniques, and the instrument is being used with notable success by Hanbury Brown and his colleagues in Australia.

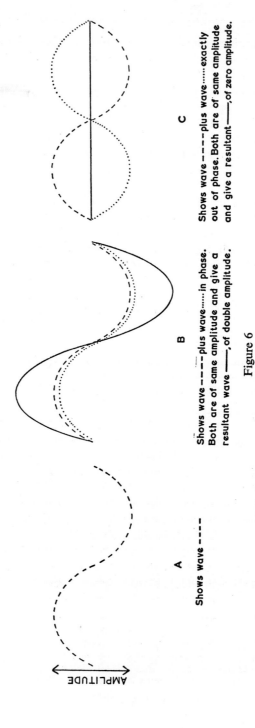

AMPLITUDE

A

Shows wave - - - -

B

Shows wave - - - - plus wave ......in phase. Both are of same amplitude and give a resultant wave ——, of double amplitude.

C

Shows wave - - - - plus wave ......exactly out of phase. Both are of same amplitude and give a resultant——, of zero amplitude.

Figure 6

The effects of waves on one another when in and out of phase. The amplitude is doubled when two waves of equal amplitude are exactly in phase, and reduced to zero when two such waves are exactly out of phase.

bands where the light waves of the two disks cancel each other, the bright bands being those positions where the light waves are reinforcing one another. Now since the stars are so far from us, their light arrives at the telescope in parallel rays, and the two images given by the two apertures may be considered to be caused by light emitted from the two halves of the star's disk (figure 7). An analysis of the optics involved – the details of which need not concern us – shows that when the separation of the apertures is adjusted so that the dark fringes disappear, then we obtain a measure of the star's diameter. The method only works for very large stars that are also comparatively close to the Earth, since only these can present a disk which, although not directly visible, is still

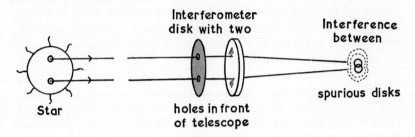

Figure 7

The principle of the stellar interferometer. Two 'spurious' disks are formed in the focal plane when the light from a distant source is divided into two separate beams which are in phase. The light from the beams interferes and, from the interference pattern and the separation of the two apertures, the diameter of the source can be measured. Hanbury Brown and his colleagues in Australia have successfully extended and enlarged the basic method illustrated here.

large enough to provide a measurable effect. Astronomers have also used the stellar interferometer for obtaining measurements of the separation of some binary stars: those where the two members of the double star system are orbiting around each other but, because of the distance of the system, appear too close together to be resolved in a telescope.

The diameters of those few stars that can be determined with the interferometer confirm the figures that have been computed from theoretical considerations of the star's radiation using a spectroscope, and this aspect of visual astronomy must now be considered further. To begin with it must be emphasized that a prism or grating will not alone prove a satisfactory spectroscope; additional lenses are required and, above all, it is necessary to restrict the incoming radiation by passing it through a narrow slit. The purpose of this is to permit greater detail to be observed in the spectrum formed by the prism or the grating, since a series of images, one in each wavelength, is obtained in this way.

It was William Wollaston who in 1802 constructed the first slit spectroscope, but the first thorough investigation of the solar spectrum made with an instrument of this kind was carried out by Joseph von Fraunhofer from 1814 onwards. Fraunhofer found that the spectrum was crossed by a host of dark lines – he mapped 576 of them – but for a long time their nature remained a mystery. Only after a number of experimenters had attempted to find an explanation was the problem solved by Gustav Kirchhoff and Robert Bunsen in 1859. These physicists discovered three important facts: first that a liquid, a solid, or a very dense gas, when heated until it glows, provided a continuous spectrum ranging from violet through to red; second, that if a tenuous gas was set glowing, it emitted light of certain wavelengths only. Their third discovery, also made independently by the physicist Balfour Stewart, was that the power of a body to emit radiation was equal to its power to absorb radiation; in the case of a tenuous gas, the wavelengths of radiation that it emitted were precisely those wavelengths of radiation that it would absorb when in the presence of hotter bodies. Moreover, it was also found by the many investigators working in the field at this time, that every chemical element emitted radiation at particular wavelengths only, these wavelengths being a characteristic of that element alone. Sodium, for instance, emits most intensely at two wavelengths in the visible spectrum: these wavelengths are 5896 Å and 5890 Å and in a spectroscope appear as two lines close together in the yellow. It also emits less intensely at other wavelengths, the relative strength of emission depending on the temperature, but it is the yellow lines that are by far the most noticeable and which give sodium vapour its characteristic yellow colour, as is so obvious in sodium street lighting.

The dark lines, or Fraunhofer lines as they are often called, could now be explained since it was clear that the Sun itself gave a continuous spectrum while layers of gases lying above the main surface or photosphere, and which could be assumed to be cooler, absorbed some of this continuum of radiation to give dark areas over the bright background. Since each element would have its own particular absorption wavelengths the dark areas appeared as dark slits or lines, and an analysis of these would enable the elements present on the Sun – or at least a proportion of them – to be determined. Incidentally, there is an interesting philosophical sidelight on this matter, for as late as 1835, twenty-five years before Kirchhoff's and Bunsen's discoveries, the philosopher Auguste Comte had published his *Cours de philosophie positive* in which he had cited the determination of the chemical composition of the stars as a perfect example of unachievable knowledge; perhaps the most notable instance of philosophical pontification limiting the technical advances that can ever be made.

The Sun, by far the nearest star to us, is an ideal object for study since it provides light and other radiations in plenty for analysis, and astronomers were not slow to take advantage of this. Coupled with experimental work in the laboratory, it was soon found that the lines themselves sometimes appeared split into components, all of which are polarized in a particular way. It was the Dutch physicist Pieter Zeeman who in 1896 found that this phenomenon was due to the radiating element being in the presence of a strong magnetic field and, furthermore, that the nature of the polarization provided evidence of the direction in which the field lay. Almost twenty years later a German physicist, Johannes Stark, observed an analogous effect caused by the presence of an electric field. Again, since elements present spectra that are different in detail at different temperatures and pressures, the close examination of spectra can provide many details about physical condition that are unavailable in any other way. In the Sun these details can be readily observed since there is enough light to permit of a very large dispersion, so that a magnified picture of the spectrum can be obtained where 4 Å cover as much as 1 mm of the photographic plate; using a microscope for examining the plates it is therefore possible to observe changes in line widths and other details that are only fractions of an Ångström.

As the Sun is near enough to present a disk, details of its surface may be studied in a way that is not possible with any other star. Straightforward telescopic observation is one obvious way, but there are also other means making use of the spectroscope or a derivative of it. If the slit of a spectroscope is directed towards the Sun's limb and is then opened wider than normal, observations can still be made at a particular wavelength, but now the large flame-like protruberances known as prominences may be observed, since a particular hydrogen wavelength is their prominent radiation and they show up well against the dark background of the sky – dark because it contains no direct hydrogen wavelength illumination. To observe the disk of the Sun in the light of a particular wavelength is not so simple, however, since any detail at one wavelength will be obliterated by the radiation from neighbouring wavelengths that will come through the widened slit. The slit must therefore be reduced to its normal size but then, of course, the area of the Sun's disk that may be observed at any one time is very small. This can be overcome by making the slit move across or scan the Sun's disk, but then the image at the other end of the spectrascope will shift and means of compensating for this must be introduced. The solution, devised independently by George Hale and Henri Deslandres in 1890, was to use two slits, one scanning the disk of the Sun and the other following the moving image. It can then either be arranged to observe visually with the two slits oscillating quickly across the field of view – the instrument is then called a spectrohelioscope – or to photograph the result, in which case the slits can move more slowly and do not need to repeat their movements; we then have a spectroheliograph.*

More recently new developments have considerably extended the field of solar observing and, in particular, made it possible to detect the outer solar corona in full daylight. Ordinarily this is impossible since the thin gases forming the outer corona are too dim to be properly observed through the light from the Sun scattered by the terrestrial atmosphere – they are, for instance, only about one thousandth as bright as the Sun's disk, the scattered light from which is also of this value near the Sun's edge. Before 1930 the only way to observe the corona was during the few minutes when the Sun becomes totally eclipsed by the Moon's disk, but in that year the French astronomer Bernard Lyot designed a special telescope that minimised the problems caused by scattered light with an efficiency never achieved before and in which an occulting disk was used to cause an artificial eclipse. First operated at a height of some 10,000 feet from the Pic du Midi observatory in the Hautes-Pyrenées, he successfully observed the outer corona and photographed its spectrum. In 1933 he also devised special monochromatic filters that, like his coronagraph, have come into widespread use. The filters have been based on interference techniques which result in a cancellation of all wavelengths but those required, and although expensive, delicate, and requiring temperature control, they are highly selective; whereas an ordinary coloured glass or gelatin filter will pass at least 50 Å, the Lyot filter can isolate a band of transmission as narrow as 0·7 Å.

In stellar observing, other improvements have been made. Another French astronomer, André Lallemand, has designed an 'electronic' telescope which is essentially an image converter, since it takes light from an ordinary telescope and then amplifies this so that the available energy

* This multiplicity of names for a device can assume almost absurd proportions, and a spectroheliograph used for taking moving pictures instead of still photographs is sometimes seriously referred to as a 'spectroheliokinematograph'.

is greater, and short exposure photographs may be taken for subjects which normally require a long exposure. This brings great advantages, as the turbulence of the air, even on a night which to the casual observer appears utterly still, blurs long exposures by causing apparent movement of the images, and the shorter an exposure can be the better. Other developments in photographic techniques have also been important, and a method devised by Fritz Zwicky at the Mount Wilson and Palomar Observatories in California has made it possible to make rapid determinations of the distribution of stars of different colours in a selected area of the sky by marrying together positive and negative photographs taken in light of different colours. Electronic techniques are also being increasingly used, not only in image converters like the Lallemand camera, but for analysing observational results, assessing stellar magnitudes and detecting variable stars whose output of light varies just sufficiently to be photographed. Attempts are even being made to devise a camera that can integrate a series of exposures taken only during moments of 'good seeing' when the atmosphere temporarily appears clear and steady, but even if such a device should come to fruition, this would overcome only one of the limitations to observing in visible wavelengths that the terrestrial atmosphere causes.

The effects of the terrestrial atmosphere on straightforward observing are greater than might at first seem possible. On the one hand the atmosphere keeps altering in transparency and, on the other, in its refracting power, the latter causing an image to move in the field of view of the telescope and, without a telescope, to give rise to twinkling. Telescopically this means that no large telescope can reach its maximum power of resolution, since an oscillating image spreads over too large an area. A telescope with excellent optics that could resolve objects as close together as, say 0·05″ in theory, will have this reduced by a factor of eight or ten times owing to atmospheric turbulence. Even if image converters enable energy amplification to provide comparatively short exposures, it is highly doubtful whether the resolution will improve by more than a factor of two. The varying transparency also provides a limit to the dimness of objects that can be photographed, although it is not the only factor that exercises an effect in this way. The night sky is not black, in spite of how it appears to the eye, because there is always a very low intensity background glow due to the bombardment of the air by electrified atomic particles from space. The glow occurs at a height between forty-seven and seventy-five miles and on rare occasions has even been known to exceed the Milky Way in brightness, although usually it is far less than this; all the same its intensity is sufficient to limit long exposures to hours instead of many nights, and so the space-penetrating power of the telescope is reduced.

It would be bad enough if these limitations were all that the terrestrial atmosphere causes but, in addition, the transparency appreciably limits the range of wavelengths that may be received at Earth-based observatories. This is advantageous from the human point of view, since X-rays and short ultra-violet radiation would cause many dangerous and indeed lethal effects should they reach the ground, but astronomically the limitations are a nuisance. The shortest wavelength that can reach us is in the ultra-violet at 2900 Å, but since the range of ultra-violet goes from 3900 Å down as low as 140 Å it is evident that only about twenty-five per cent of the total ultra-violet radiation incident from space can be observed. The physiological effect of this twenty-five per cent is serious enough, as evinced by the pigmentation we call sunburn that the body generates to prevent penetration to lower levels of these short wavelengths, but for the spectroscopist this is nowhere near good enough. The Sun and many other stars radiate strongly in the ultra-violet, and particularly in the Lyman series of wavelengths, of which the longest (the Lyman α or Lyman-alpha line) is only 1216 Å, and thus the atmosphere prevents this radiation and, more especially, its spectrum from being observed at all. Theodore Lyman discovered his series of ultra-violet lines in the laboratory in 1914, but although their presence was theoretically expected in the solar spectrum, it is only in recent years that they have actually been observed, as will be discussed in Chapter 8

The major part of the ultra-violet is absorbed by the oxygen present in the atmosphere, but the water vapour and carbon dioxide also play their part in filtering away radiation, although they operate at the long wavelength and from 11,000 Å (about 10^{-3} mm) to wavelengths of about 1 mm, in other words including much of the infra-red, the total range of which extends, as we have seen earlier, from 7500 Å to 300,000 Å (0·1 mm). Unfortunately the absorption does not then cease and the 'window' that the atmosphere gives from wavelengths of 0·1 mm onwards does not extend indefinitely, and it closes down again for wavelengths greater than ten metres. This last absorption range, which seems to extend without limit to the longest wavelengths, is caused by the ionosphere – a cloud of atomic particles released from electrified layers of air – that acts as a radio mirror, reflecting back into space all radio wavelengths that reach the Earth except for those in the 0·1 mm to 100 metre band. The ionosphere itself is known to be caused by short wavelength radiation from space impinging on the air and, naturally enough, its very existence acts as a stimulant to the astronomer to discover the

sources of this radiation. To do this with any pretentions of complete-
ness means making observations above the atmosphere, but the foundations on which such extra-terrestrial measurements are based are
more conveniently made from Earth-based observatories. After all, some
infra-red and ultra-violet radiation is unaffected by the hazards of
passing through the atmosphere, especially if the observatory is situated
high enough and is sufficiently isolated to be above the densest and away
from the dirtiest atmospheric conditions. This is the case in observatories like those at Mount Wilson and Mount Palomar in California,
Mount Stromlo in Australia and the Union Observatory at Johannes-
burg, to name only a few, while some observatories such as the Pic du
Midi already mentioned, and the Climax Observatory at Boulder,
Colorado, lie above 9000 feet and are eminently suited for observations
with the coronagraph and into the shorter range infra-red and longer
wavelengths of the ultra-violet.

Observations made at Earth-bound observatories are still vitally
important and, in radio astronomy, provide the one means of identifying
radio sources with objects that can be recognized – at least as far as
their general classification is concerned. They are also important in
their own right, and it would be foolish to abandon the visual spectrum
because we can now probe into the many invisible sections. The whole
general picture of the universe and the measurements made in it is based
on observations from optical observatories – this is the background
against which all the new facts must be set. Before examining the extreme
ranges of the spectrum it is, therefore, necessary to assess the information that optical Earth-based observatories have obtained, and then
sketch the kind of universe that arises from them. This will occupy the
next two chapters, but with this background it will then be possible to
place the new evidence from invisible astronomy in its proper context.

CHAPTER 3

Visual Observation

With an idea of the visual spectrum in mind, it will be clear that there are two basic kinds of observation that the optical astronomer can make – observations in integrated (white) light and observations in selected wavelengths. Integrated light is used for measurements of position and motion, and for most magnitude determinations, while selected wavelengths are utilized for specialized studies of the chemical composition of celestial bodies, the analysis of stellar conditions and for obtaining clues about their internal constitution and behaviour. With this raw material to hand, a picture of the universe can be prepared, even though it will be incomplete (since only some of the total radiation information is being used); moreover, as there are so many problems still to be solved, the picture to be drawn will be far from final.

It will be convenient to begin with one important aspect of astronomical measurement – the question of distance determination, which is fraught with more difficulties and far more uncertainty than is usually realized. The most straightforward measures of distance, and those on which some reliance may be placed, are made by a method analogous to that adopted by terrestrial surveyors when they want to determine the distance of an inaccessible object. Consider, for instance, the problem of finding the distance of the point C from the point A (figure 8). If C is inaccessible, then it is not possible to use a measuring tape and optical means must be employed.* The observer therefore measures the angular distance between C and some distant fixed point D: he obtains the angle DAC. Now the observer moves to another point B, and carefully measures the distance AB. A second series of observations is next made to obtain a new separation between D and C; this gives the angle D. The application of trigonometry then allows half the angle BCA to be determined – the so-called 'parallax' of C – and a little trigonometry permits of the calculation of the distance between A and C.

* As with all analogies, the statements should not be taken too literally. In modern surveying there are new devices that can measure the distance of inaccessible objects directly by emitting beams of light or pulses of radio waves and timing them as they travel from A to C and back again, but for our purpose these are irrelevant – at least at this stage of the argument.

IIIA. A cluster of galaxies in the constellation of Hercules. The stars in the photograph are part of our own Galaxy and one may think of them as a curtain through which the distant galaxies are viewed. Spiral, barred-spiral and elliptical galaxies can be seen; the most notable are the pair of spirals (*centre*) and barred-spiral (*right*) just below the middle of the photograph. The distance of the cluster is some 340 million light years. Photographed with the 200-inch telescope.

IIIв. A pair of spiral galaxies (NGC 5432 and 5435) close to one another in space. A close examination of the photograph shows that there are bridges of intergalactic matter between the spirals, leaving no doubt about their physical connection.

The astronomical method of direct distance measurement is similar, and is known as 'trigonometrical parallax'. Clearly, the greater the separation between A and B, the larger the angle BCA for any given distance of C, and since the stars are so distant – early failures to measure the parallax of any star at all prove this beyond doubt – astronomers have adopted the maximum separation they can. They make observations six months apart, so that A lies on one side of the Earth's orbit, and B on the other; the average distance of the Sun is ninety-three million miles and therefore this bi-annual observing method provides a base-line between A and B of 186 million miles. A similar method, where A and B are two widely separated observatories on Earth, can, in theory, be used for obtaining the Sun's distance, but in practice the errors are too great in view of the very limited possible separation of A and B.

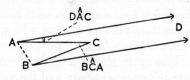

Figure 8

The principle of measuring the distance of an inaccessible point (C). An observer at A determines the angle of C using a distant reference point D (this gives the angle DAC). The observer next moves over a measured distance to B, and again measures the angle between C and the distant reference D. By trigonometry he can then determine the angle BCA and thence the distance from the mid-point of AB to C.

Other methods have to be used, and one of the most successful has been to measure the comparatively small distance of a minor planet (planetoid) when it approaches close to the Earth on its eccentric orbit round the Sun: gravitational computations of its orbit and the orbit of the Earth then provide a value for the solar distance. There is a small number of planetoids that make approaches closer than twenty million miles and as a result the distance of the Sun is now known to an accuracy of about one part in nine thousand – in other words over the ninety-three million miles to the Sun the error is no greater than 11,000 miles.

With the solar distance known, using the most refined methods of observation, and making due allowance for all possible errors, trigonometrical parallax can provide a measure of distance up to some 325 light years,* but beyond this the probable errors of observation become

* The light year is a measure of distance, not of time, and is that distance over which light will travel in one year. It is 6,000,000,000,000 miles (6×10^{12} miles). The distance of the nearest star α Centauri is 4·2 light years.

too great for the results to be reliable and other methods have to be found. Of these the most familiar is that which makes use of the fact that the Sun is moving in space. Since, as Edmond Halley discovered in 1718, the stars actually possess motions of their own, it seemed likely that the Sun also moved in space, yet no direct proof was possible for a long time. Only in 1783 could William Herschel make an attempt to deduce this motion from an analysis of the movements of other stars. His argument was that if one observed in a direction in space towards which the Sun was moving, then the stars would all appear to be moving away from that part of the sky, just as objects appear to move to the left and the right from a point directly ahead when one is driving in a car. This point Herschel called the apex, and the point from which the Sun was moving, the ant-apex: the ant-apex could, he thought, be recognized as the apparent place in the sky towards which the stars appeared to be moving. In 1783 comparatively few stellar motions were known, but using his hypothesis and appreciating the fact that stars lying at the 'side' (i.e. between apex and ant-apex) would show the greatest effect due to the Sun's motion, as objects at the side of a speeding vehicle do, Herschel was able to compute the apex, ant-apex and the velocity of the Sun. From a statistical point of view the few stars for which motions were available – some nineteen – was too small a sample; in 1805 he was in a position to use an additional forty but, as it turned out, his results were very similar in both cases. Later determinations have not substantially altered the positions he found, although it has become clear that different values depend upon the class of celestial objects one uses as a basis for the calculations. For instance, if bright blue stars are taken, these give one value for the position, clusters of stars give another value, and so on; yet in spite of the discrepancies the solar motion can be used as a means for determining stellar parallax.

In principle the Sun's motion can provide parallax measurements because over the years the Sun moves a considerable distance in space, a distance far greater than the 186 million mile diameter of the Earth's orbit. The Sun's velocity in space is of the order of $12\frac{1}{2}$ miles per second (the precise value depending on the objects we select for determining the direction of stellar motion) and in one year it therefore travels about 400 million miles (4×10^8). If, then, one selects a star that has no apparent trigonometrical parallax, but which shows a parallactic shift against the background of still more distant stars after a period of a few years, it is possible in theory to calculate its parallax due to the Sun's motion since one has, for example, a base-line of 12×10^8 miles after a three-year interval between observations. As the Sun's motion continues over a long period this method is usually known as 'secular' (i.e. age-long) parallax but it has a grave disadvantage. This arises since the star whose distance is being determined will undoubtedly have its own motion, and it is impossible in the case of one isolated star to separate its individual motion and its apparent shift due to the solar motion. However, if we determine the parallactic shift for a number of stars, preferably all of similar type, then, if the sample of stars is large enough, we may assume as many will be moving in one direction as another – in other words that on the average there will be as many stars moving in the opposite direction to the solar motion as in the same direction – and their individual motions should cancel out one another. Of course this argument is only statistically true, and in any group of stars it will not be precisely so: all the same, the errors arising from this assumption will be small, and the statistical parallax obtained will possess some validity and will enable distances to be determined for stars further away than those for which parallax can be found trigonometrically. Statistical parallax permits distance determination out to some 1600 light-years.

Many stars – about half those in our own Galaxy – are multiple systems in which one or more stars are in orbit around a common centre of gravity. Some can be observed visually to be behaving in this way, as William Herschel found; some are discovered to be orbiting because as they pass in front of one another their total light varies in a specific way; still others may be detected by an interferometer (as mentioned in Chapter 2), and there are those which are too close together to be directly seen, but whose spectra show a regular series of Doppler shifts that give the clue to their multiple nature. If a pair of stars is in orbit and observations are made of them over a sufficient period, it is possible to deduce their parallax, and this can prove to be a very useful method of distance measurement. The procedure for determining such a 'dynamical parallax' is based on the gravitational behaviour of bodies in orbit, for observations readily provide the period for one complete orbit of one star with reference to the other, and the apparent separation (in seconds of arc) of the two stars. If, then, the masses of the stars are known, the real separation (in miles) may be computed and so the parallax determined. The difficulty is that to determine the masses exactly, the parallax must be known, and this is the very thing we want to discover, so once again it proves necessary to make an assumption. Now although stars vary widely in their intrinsic brightness and also in their size, their masses do not, in fact, spread over an equally wide range – at least for the majority. An investigation of the masses of binary stars in systems where the parallax of the pair of stars is already known by some other means, makes it clear that for such systems the range of masses is small – from 0·15 to 3·5 times the mass of the Sun or a factor some twenty-three times – whereas the luminosities vary by a factor of 25,000. Moreover, the formula that is computed so that the parallax may be derived from the dynamics of a binary system, varies only as the cube root of the masses of the stars concerned; even if the error is as great, therefore, as twenty-three in the assessment of the mass, the error introduced into the parallax will only be 2·8 (the cube root of twenty-three). But the masses are unlikely to be as wrong as this, and to begin with it is generally assumed that the masses are each equal to the Sun – an assumption that is made far more likely if only stars in the main sequence* are chosen. With this assumption, the parallax of the system is then determined, and the value obtained permits a calculation to be made of the intrinsic luminosities of the stars. Since there is a relationship between the mass and the luminosity of the stars in a binary system, new values can be obtained for the masses and these values inserted back in the formula to provide a new and improved value for the parallax. By this method, it is possible to reduce errors from the maximum twenty-eight per cent given above to a maximum of fifteen per cent, with five per cent the normal value for the uncertainty.

Only visual binaries may be used for determining dynamic parallaxes, but this still does not permit distance measurement as far as one would wish, and other indirect methods must be adopted even though the uncertainty will increase the more stages of assumption there are in the process, however good one's guesses may be. Spectroscopic parallax methods are widely used, since they can be applied to more distant objects than any of the techniques so far mentioned, and really depend only on the star being near enough, or bright enough, to give a good clear spectrum that can be studied to provide an assessment of the class of star that is being observed.

Stars may be classified by their spectra, and it was Angelo Secchi who in 1867 first proposed such a method. Secchi's system has been much modified in the century that has followed his work, and is now based not only on the observational differences but also on theoretical evidence that has been derived from studies of the behaviour of atoms. The system of classification now used is based on the work of Edward C. Pickering of Harvard College Observatory and Annie J. Cannon: it

---

* Main sequence stars will shortly be discussed – here it is only necessary to note that the majority of stars are of this type.

resulted in the publication of the Henry Draper* Catalogue of stellar spectra which consists of nine volumes and was completed in 1924. This contains seven main spectral classes, each denoted by a letter of the alphabet, and with each letter interval divided into ten subdivisions; to it have been added four other classes. Because of the changes over the years while the system was being developed, the letter classes do not run in alphabetical order, and the entire series now reads Wolf-Rayet, O,B,A,F,G,K,M,R,N,S.† The first, the Wolf-Rayet stars, are named after C. Wolf and G. Rayet, two Paris astronomers who in 1867 discovered the existence of some bright blue stars, with a few very bright-lines in their spectra and which have later proved to possess a central core surrounded by an expanding shell of gas: they lie at one end of the rest of the sequence of letters which present a continuous 'development' in spectra. The O stars contain some weak bright lines but are noteworthy for lines caused by ionized‡ helium gas and ordinary helium and hydrogen; they are blue-white in colour. The B stars show no ionized helium, but some ordinary helium is present in most of them (B0 to B9), hydrogen is more strongly in evidence than is the case with the O-type stars, and ionized vapours of metals are present. In A-type stars the hydrogen lines are extremely strong – more so than in O or B – and as one passes from A0 to A9, the lines caused by ionized calcium vapours increase in intensity. This strengthening of ionized calcium increases still further in F-type stars with a diminution in the intensity of hydrogen. This change continues in the G class, of which the Sun is a typical member (G2), with the vapours of normal (non-ionized) metals also beginning to show, a characteristic that becomes stronger in the following K-type stars in which hydrogen lines are almost completely absent. The M-type present dark bands in their spectra and these become stronger from M0 to M9: they are due to the presence of titanium oxide. Stars classified as types R and N display bands also due to compounds, but in this case they are carbon compounds with, in the N-type, lines such as those of the metal potassium which has a relatively low boiling point. The S-type stars also present bands due to compounds, but not only of titanium oxide as in class M, for here zirconium oxide is also present.

The stars in this classification provide differences in chemical composition – at least in their outer gases where the dark lines are generated

* Named in memory of the pioneer spectroscopist Henry Draper who died in 1882 and whose instruments and a sum of money were left to the Harvard Observatory.

† There is a well-known mnemonic for the letter series 'Oh, be a fine girl, kiss me right now, smack' which is effective if not edifying!

‡ An ionized gas (to be considered more fully in Chapter 5) is one in which the atoms have lost an outer particle and so possess a positive electric charge.

– but they also differ in colour. O and B type stars are bluish white, A stars are white in colour, F have a creamy colour, G are yellow, K orange and the four types M,R,N and S are red. That is not to say that they fail to have a continuous spectrum of all colours against which the dark lines are displayed, but that in each and every class the strongest intensity of the background continuum is different. Now it must be obvious that such a classification is likely to have some underlying cause, and both the colour and the presence or absence of ionized vapours give one simple answer – the classification is one of temperature. What the physicist calls a 'black body' – one that appears completely black when cold, since it then absorbs equally all wavelengths of radiation – will radiate in a colour that depends on its temperature. An iron poker, for example, although differing a little from the perfect theoretrical black body, is a dull red when heated; then, if heated further, becomes a brighter red, changes to a straw colour and, if heated until molten, will emit a 'white heat'. Stars approximate to black bodies and the colour they display is an indication of their temperature. So too are their spectra, for the presence of compounds (titanium oxide, carbon compounds and so on) indicates a comparatively low temperature – otherwise the compounds would be broken down into their separate constituents; the vapours of metals is sure evidence of a higher temperature, while the appearance of lines due to electrified substances shows that the temperature is higher still, for in these cases the atoms composing the substances are electrified or 'ionized' and this can only be achieved by a considerable input of energy, especially in the case of helium.

The spectral classification therefore provides evidence about stellar temperatures, and the range of these is considerable. Stars of class O5 have their surface gases as hot as 36,300°K,* by B0 this has dropped a little to 26,600°K and so on, downwards, the Sun having a temperature of some 6000°K (5727°C), with the M stars and other low temperature stars at anything between 3400°K and 2600°K.

From such a classification as this it might be imagined that some guess can be made of the intrinsic luminosity of a star since, if it radiates like a black body, one would expect it to become brighter (as a poker does) the hotter it is. And, of course, if intrinsic brightness can be deter-

---

* K stands for degrees Kelvin, named after Lord Kelvin whose work on heat in the nineteenth century led to the establishment of a temperature scale in which zero degrees is that temperature at which a gas would cease to exert any pressure at all, due to its molecules (groups of atoms) having insufficient energy. Experimentally this absolute zero is unattainable, although physicists have come to within a fraction of a degree of it. Compared with the better known Centigrade or Celsius scale on which zero is the freezing point of ice, absolute zero is—273·2° on Kelvin's scale.

mined then, as it is not very hard to ascertain the apparent brightness of a star, one can compare the two and from this comparison compute the star's distance. One has to make the assumption that radiation decreases in a particular way with distance, and what is done is to assume that this decrease is the same as that determined in the laboratory. This fits in well with theory, and is sometimes referred to as an inverse square law since the radiation diminishes by a factor depending on the square of the distance.* However, although this is admirable in principle, it is far more complicated in practice because the yellow and red stars (half-way along the F range down to and including M) may be of two kinds, one of intrinsically low total luminosity and the other intrinsically high. These two sub-types are called dwarfs and giants, the terms (meaning dwarf or small in intensity, and giants in brightness) originally being used by Ejnar Hertzsprung and Henry Norris Russell in Princeton but only as an indication of intrinsic brightness and not of size. It later turned out that size was also involved, since if we have a star at a certain temperature then, from what has been discussed so far, it must be radiating at a certain rate. A square centimetre of the star's surface will radiate so much energy and no more. The intrinsically faint star will be faint not because its radiation per square centimetre is low but because it does not possess many square centimetres – in other words, because it is small. Conversely, the intrinsically bright star of the same spectral class will, again, be radiating in the same way and is only bright because it has more square centimetres from which to emit radiation – it is, in fact, large. Hertzsprung and Russell obtained these results from spectroscopic observations coupled with parallax determinations, and with the evidence of other observers were able to draw up what has now become a famous diagram – the H-R diagram – which depicts the relationship between spectral class and intrinsic luminosity. A simplified H-R diagram is given in figure 9 (p. 57), the spectral classes being shown along the bottom and the intrinsic luminosity – absolute magnitude† – on the left. The position of our Sun is clearly indicated, and some other well-known stars are also shown by name.

It will clearly be seen that from about F5 downwards there are the two

* Simply if d is the distance, then the intensity I is proportional to $1/d^2$.

† Ordinarily the brightness of a star is given in magnitudes, but these are *apparent* magnitudes since they define only the apparent brightness of a star. The apparent brightness depends on the star's distance as well as its brightness, and an intrinsically bright star may well appear of the same magnitude as an intrinsically dim one if the dim one is much nearer. Absolute magnitude is defined as that magnitude which a star would appear to have if situated at a distance of 32·6 light years (a figure chosen because at this distance a star would have a trigonometrical parallax of 0·1″, and this makes for easy calculation).

brightness classes of dwarfs and giants. The dwarfs are not very different from the Sun in size, some having diameters larger by a factor of 1·5 or 2·0 while others are smaller by a factor of, say, 0·5; on this evidence the Sun itself is a dwarf. Giant stars may be very much larger than the Sun and values of some 16 to 130 times are not unusual. In addition, as the H-R diagram shows, there are some 'supergiant' stars, the sizes of which are really immense compared with the Sun. The star Antares, for example, has a diameter 300 times that of the Sun, one in the constellation of Cepheus (VV Cephei) is 1200 times larger and the star ε (Epsilon) Aurigae has a diameter 2000 times the Sun's, which means that if placed where the Sun is, it would swallow up most of the solar system, leaving only the planets Uranus, Neptune and Pluto beyond its boundaries.

With this knowledge to hand, it becomes practicable to obtain parallax values from measurement of apparent magnitude and the computation of absolute magnitude based on the spectroscopic evidence. Such 'spectroscopic parallaxes', with evidence adduced about detailed spectral differences between giants and dwarfs as investigated by three American astronomers, W. W. Morgan, P. C. Keenan and E. Kellman, possess an uncertainty of some fifteen per cent, but for distant stars this is better than any other method at present in use. However, spectra can give another clue to distance and this arises from the presence of interstellar dust. The dust, which seems to consist primarily of hydrogen compounds with a great number of ice crystals and probably some iron oxides and carbon as well, is found in very small particles, or so it is conjectured from an analysis of its differing ability to absorb radiation at various wavelengths. The dust is sparse – there is about a quarter of an ounce per volume of a billion cubic light years – yet its particles have so small a size that a volume of only one cubic light year would contain between 10^6 and 10^{10} particles, which is a very large number and one that will give a considerable absorption of radiation over many light years of space. The dust dims the light of distant stars and so makes their apparent magnitude smaller (numerically greater) than it would be if space were empty, and some allowance must be made for this in the determination of spectroscopic parallaxes. An examination of spectra shows that they are crossed by dark lines that do not show any Doppler shift as do the lines originating in the stars, and these are the lines due to interstellar material. The thickness of a line is a measure of the concentration of the dust and, with a figure for its density in interstellar space, the distance through which starlight must have passed to give rise to this intensity may be calculated. This can provide yet another clue to stellar distance, but its reliability is not very high since there is reason to believe that the dust is not uniformly distributed in space but, rather, that it is to be found in patches.

So far, then, it seems that there are four main methods of determining astronomical distances – trigonometrical, statistical, dynamical and spectroscopic parallax techniques. Yet the possibilities are not exhausted, at least as far as some special objects are concerned. Some stars are explosive, ejecting a spherically shaped shell of gas. In the sky they are noticeable because they suddenly appear visible to the unaided eye in a place where no star has previously been observed. This is merely a consequence of their increase in brightness by a factor of some 100,000 times, although to anyone unfamiliar with the cause, a new star appears to have sprung into life. For this reason the stars are called 'novae' and their distances are determined from observations of the expanding shell of gas. Two observations are required, one looking directly at the star, and the other at the edge of the shell. On looking directly at the star with a spectroscope some of the lines will be due to the shell of gas, and, since this is expanding, will display a Doppler shift. From this the velocity of the gas directly towards the observer may be calculated. The observation made at the edge of the shell will determine its apparent distance (in seconds of arc) from the central star, and another observation made months later will provide a new value for this apparent distance due to the continued expansion of the shell. From the edge observations, the shell will be found to have moved a certain number of seconds of arc during the interval, but as the spectroscopic measures give the actual velocity of the expanding gases, the edge movement can be translated into miles: thus the scale, miles to seconds of arc, can be determined, and thence the distance at which the nova must be to provide this scale. The number of novae is comparatively small and this method of distance determination is therefore limited, useful though it can be on occasions.

Novae may be classed as one type of variable star, particularly since some eject shells at regular intervals: those so far observed to behave in this cyclic fashion have periods of some thirty to almost eighty years. The majority of variable stars are less dramatic in appearance, varying their intensity by no more than six magnitudes at the most, and usually by far less. Some vary irregularly, undergoing slow and, at present, completely unpredictable changes, whilst others vary in a semi-regular fashion, the period fluctuating about an average value. Of the regularly varying stars, two kinds are recognized: those with periods of sixty days or more and those with periods less than this. Lastly there are flare stars and supernovae. Flare stars show very rapid luminosity changes that appear suddenly and last no more than a few minutes or half an hour at the most; all are red dwarfs. Supernovae display immense changes of magnitude, far in excess of those of the ordinary novae, which are themselves surprising enough; whereas a nova may alter in brightness only a few hundred thousand times at most, the supernova will change by anything up to a billion times. A supernova is a prodigious explosion in which a star loses some one per cent to ten per cent of its total mass and ejects this as a gaseous shell that expands with a velocity that may reach 7500 miles per second. Such explosions as this are rare and appear to occur on the average once every 300 to 400 years in any particular galaxy: of the fifty or so observed, only four have been in our own Galaxy and these were all in historic times. As will be clear shortly, the supernova can be used as a means of measuring very great distances.

The ordinary regular variable star is not adaptable for distance determination, but matters are different when we turn to three types of short-period variable – the Cepheids, the RR Lyrae and the W Virginis variables, all named after particular stars typical of their class. The Cepheids have periods of variation ranging from one to fifty days, with the majority showing around a five-day fluctuation. The star δ (Delta) Cephei, the first example of this variable to be discovered, displays a relatively fast increase in brightness and a slower decline, and a change in spectral class during this variation which covers no more than about one magnitude. All are yellow giants or supergiants. The most important factor about stars of this kind is the relationship of period of variability with intrinsic brightness, for in most cases the longer the period the brighter the star. The W Virginis type has lower brightness than a Cepheid of similar period, but RR Lyrae variables have a different relationship. The Cepheids lie all around our own Galaxy but the RR Lyrae type is found mainly in open clusters of stars, and most of the W Virginis in globular clusters. The open clusters are aggregates of stars irregularly arranged and spread over anything from five to fifty light years in a flattish disk within the centre of the Galaxy; they each contain some 15 to 300 or more stars. The globular clusters, on the other hand, contain tens of thousands or even millions of separate stars, yet their diameters are found to be no more than 400 light years, and some are spread over as little as twenty-three light years; they are found to be distributed in a kind of spherical ball that forms what can best be described as a halo to our Galaxy.

Clearly, since the period of fluctuation of such stars forms a definite relationship with their intrinsic brightness, they serve admirably for distance determination and are widely used for this purpose. What is more, stars of this kind have been observed in nearby galaxies and, with the odd supernova explosion, provide a convenient way of determining the distance of such objects, for galaxies are so far away that they present many problems when it comes to analysing their distribution in depth. To begin with, there are a number of different kinds of galaxy. Our own Galaxy, it is now certain, is a flattish disk with a central bulge and spiral arms. It is a spiral galaxy and its size, determined by using all the means so far described, is about 81,000 light years in diameter and some 13,000 light years thick at the centre, although it possesses a halo of globular clusters, RR Lyrae variables and some other stars of spectral classes of about A7 down to M. This spherical halo has a diameter of around 98,000 light years.

Our Galaxy contains some 10^{11} stars, and includes not only stars and dust but also a considerable amount of gas, some of which is dark and some bright. The gaseous clouds, known as 'nebulae', are ordinarily dark but some appear bright because of the radiation from stars embedded in them. Their spectra show bright lines only, which is the reason that their gaseous nature is known.

This mixture of stars, dust and gas is typical of a spiral galaxy, of which many millions have been photographed. It is typical also of the content of what are termed 'irregular' galaxies, that is galaxies which are amorphous in shape, a typical member of which is the so-called Small Magellanic Cloud, visible as a hazy patch of light in the skies seen from the southern hemisphere, and similar in appearance to the hazy band of light – the Milky Way – that stretches over the skies seen from both hemispheres and which is part of our own Galaxy. But there is another class of galaxy that has been observed in its millions, and this is known as the elliptical, since many members of this class are elliptical in shape. Such galaxies present no structural detail as does a spiral, except that their central parts are more concentrated than their outer parts. It appears that an elliptical galaxy contains mainly stars and little, if any, clouds of free gas, and their sizes seem to be smaller than the spirals by a factor of 0·6.

In order to formulate a picture of the universe some estimate must be made of the distances of galaxies, yet this proves to be difficult because most are certainly very far away and the more usual methods of distance determination break down. Clearly trigonometrical parallaxes can be of no use – they cover only stars comparatively close to the Sun and cannot deal with most objects within our own Galaxy, let alone outside. Statistical parallaxes are again inapplicable since they are derived for stars showing a parallax due to the motion of the Sun; dynamical parallax is unsuitable as we are not here concerned with binary star systems but galaxies, and spectroscopic parallaxes cannot be used because it is not possible to analyse the spectra of separate stars in a galaxy. For the nearer galaxies, the use of Cepheid and allied variables can be a very important method. Again, for such galaxies, the rare supernova outburst, the recognition of other objects such as globular clusters, supergiants, open clusters and objects known as planetary nebulae (extended oval masses of gas with a hot bright star at the centre) may help, as all have an intrinsic brightness that is known with sufficient precision. The difference between apparent magnitude and absolute magnitude provides a distance measure, technically known as the 'distance modulus'. Nevertheless it is not long before every one of these methods fails because the distances become too great, and for galaxies further away than 16 million light years some other means must be found if we are not to remain content with 'very far off'.

One obvious method is to determine, from those galaxies whose distances are known, a figure for their absolute magnitudes. Using this we then have an average brightness for a particular type of galaxy, but the method is not completely satisfactory because not all galaxies of the same type are equal in brightness. However, if a figure of average absolute magnitude of -15 is taken, and if due allowance is made for the type of galaxy being observed, approximate values of distance may be obtained for galaxies out to some 325 million light years; after this errors will become too great due to probable radiation absorption and difficulties in determination of the type of galaxy.

Another method that has been adopted is to measure the apparent angular size of galaxies and then, assuming that galaxies of a particular type have an average size which can be computed statistically, use this size to compute their distance. This assumes that galaxies appear smaller the further away they are, an assumption that seems straightforward enough and unexceptionable since terrestrially, as well as in the nearer reaches of space, we find just this. Yet one cannot be certain that such an assumption is true once measurements are taken further than local (within the Galaxy) distances, and if it seems churlish to question so apparently fundamental and obvious a fact of experience, it may be noted that in the mathematical model of the universe computed by Albert Einstein and W. de Sitter, there is a particular distance at which a galaxy has a minimum apparent size, beyond which it begins to appear larger again. Of course, this model of the universe is not necessarily correct, but it does underlie the fact mentioned in the first chapter, that we should be aware of the assumptions made at every stage of our argument, and that their apparent common-sense should not blind us to the fact that what applies in one situation may not be strictly applicable when matters are taken to extremes.

The Einstein-de Sitter change of size phenomenon also brings to the fore the whole question of distance measurement described in this chapter, from trigonometrical parallax onwards. All through it has been tacitly assumed that the shortest distance between two points is a straight line and that light travels in straight lines, yet these are statements that are correct for small terrestrial distances but may not be valid for the whole of space. Space may be of such a kind that nothing, light beams included, can travel in a straight line, but are constrained to move in some other path, and cosmologists do indeed find it convenient to use a mathematically curved space in their calculations; a space, all the same, that over very small (local) distances, reduces to the kind of Euclidean space taught at school.

Before leaving the crucial question of distance measurement, one of the basic contributions of visual astronomy, some mention must be made of another technique of distance determination, which is founded on a number of assumptions but is nevertheless a very powerful method. This is to utilize the observational evidence of the shift of spectral lines towards the red end of the spectrum which, as has already been mentioned, indicates the recession of objects away from the observer. In 1929 Edwin Hubble at Mount Wilson Observatory undertook an examination of the distances of galaxies determined by the methods described above, and found that the further off a galaxy lay, the greater the red shift of its spectral lines. For galaxies too far off to have their distances determined in any other way, the red-shift was measured; then, based upon a simple proportional relationship between distance and shift that Hubble had formulated, it could be used for computing the distance of any galaxy. This relationship, known as Hubble's Law, breaks down when the distances are very great, but the situation may be saved by the application of Einstein's theory of special relativity, which takes into account the correction required when relative velocities – in this case between our Galaxy and the most distant galaxies – are very large. The implications that follow from these measures will be discussed in the next chapter.

So far the observations that the visual astronomer makes have been concerned with measurement of distance, of motion, of brightness and of spectra. These form the bedrock of astronomical knowledge, but even from Earth-based observatories some excursions may be made into the ultra-violet and infra-red parts of the spectrum. These excursions cannot travel far along the full ranges of the electromagnetic spectrum due, as already stated, to the filtering effects of the terrestrial atmosphere, but that need not – and does not – prevent the astronomer from probing the spectrum as far as he can. Special adaptions of techniques used for the visible ranges of the spectrum are applied to the infra-red, but for the ultra-violet a grating in the spectroscope will achieve all that can be done in the very limited range of short wavelengths that can penetrate to the ground. Ultra-violet observations have shown that the O and B type stars radiate strongly at the short-wave ends of their spectra, as might be expected with stars that are so hot, and they are also used to provide part of the analysis of the radiation of all stars. Here measurements are made of the radiation intensity in the long wave ultra-violet, the blue (to which photographic plates are most sensitive), a yellowish wavelength (where the eye is most sensitive), and in the red. Filters are used to obtain the different ranges, and nowadays photomultipliers are often employed in place of photographic plates since, as these electronic devices operate by transforming radiation into an electrical current which is then amplified, they prove more sensitive and so allow measurements to be made on very dim stars. In addition, the use of electronic devices such as these permits of the results being fed directly into a computer card punching device so that the observations may be analysed automatically. The results of these brightness measurements has led to an improved understanding of the radiation distribution at various wavelengths for stars of all spectral types.

Since the atmosphere limits the observation of ultra-violet to wavelengths longer than 2896 Å, space probes must be used for the shorter wavelengths, and this is also true of the far infra-red, where the atmosphere only gives a reasonably clear window to a range of wavelengths from about 8500 Å to somewhere in the region of 22,000 Å. Yet within this window there is a considerable amount of work that can be undertaken, while even the absorption itself, as in the case of short wavelength absorption, may be investigated to provide information about the nature and the extent of the Earth's atmospheric covering. Astronomically, infra-red observing from Earth-based observatories is concerned with two kinds of subject – planetary atmospheres and conditions, and stars. Planetary atmospheres may be examined spectroscopically, for sunlight passes through such an atmosphere, irradiates the planet, and is then reflected back through the atmosphere again, the double passage enhancing the absorption lines which the atmosphere causes. However there are severe limitations here since if the planetary atmosphere contains much water vapour or carbon compounds, the absorption that these cause will lie just at those infra-red wavelengths where the terrestrial atmosphere itself absorbs; nevertheless although mistakes have been made – absorption bands due to hydrocarbons that appeared in the spectrum of Mars were later found to be caused by absorption in our own atmosphere – some progress has been achieved. At the Dominion Astrophysical Observatory in British Columbia, the presence of oxygen was detected in the atmosphere of Venus and, in addition, the quantity was found to be very small, a fact that has recently been confirmed by the Russian space probe Venus 4 that made a soft landing on the planet in October 1967. Infra-red studies have also been initiated on the famous and beautiful ring system that surrounds Saturn, and have made it clear that while ice crystals are present in the system, these are of water ice and not, as had been previously suggested, of dry ice (frozen carbon dioxide).

The Moon has also proved a useful subject for infra-red investigation. John Saari and W. Shorthill of the Boeing Scientific Research Laboratories have made a number of scans of the lunar surface, using reflected violet light and then comparing their results with scans in the infra-red. The observations were made during the time of a total lunar eclipse when the Full Moon, having been receiving the effects of direct sunlight, is suddenly plunged into almost complete darkness on entering the Earth's shadow. Some 1000 hot spots were discovered on the lunar surface, and of these 400 were carefully plotted and identified. Most of the hot places turned out to be within the walls of craters that appear noticeably white at Full Moon, and those other areas that proved to be hot were also ones that appeared bright on these occasions. The temperature differences between these hot patches and the surrounding lunar surface were considerable, and in the crater Tycho, for instance, the floor appeared to be 48°K warmer than the surroundings, which were then at 178°K. This means that both areas were still cold since 178°K is −95°C and the floor of Tycho, although hotter, was no more than −47°C, while the temperature with full sunlight would be of the order of 100°C (373°K). There is no clear reason known at the time of writing for these hot patches, but one may assume that they are caused by a different composition of the lunar ground compared with the general surface, possibly where radiation can penetrate further than usual, thus providing a deeper layer of warm surface that takes longer to radiate away its heat; but whatever the answer, it may well have to wait until manned exploration before a full explanation is possible.

Stellar infra-red investigations have recently received encouragement by the development of sensitive lead sulphide detectors for the window wavelengths, detectors that extend the range to which photographic plates will respond. In 1937 Charles Hetzler at Yerkes Observatory showed that some stars seemed to be very cool, having temperatures not in excess of 1000°K, but it is only recent developments that have allowed this work to be followed up. Hetzler discovered what may well be termed infra-red stars, since these objects gave a radiation increase at infra-red wavelengths some nine or ten times greater than at visual wavelengths. Some of these stars appear to possess surface temperatures less than 800°K and such objects in Cygnus and Taurus have been studied carefully by G. Neugebauer, D. E. Martz and R. B. Leighton. These have a peak of radiation at 20,000 Å in one case and 50,000 Å in the other, with a very small output at wavelengths that are appreciably shorter. Their recent survey with a 62-inch plastic coated aluminium reflector shows the presence of as many infra-red as visually observable stars, and in R Monocerotis an infra-red star within a gas cloud – possibly a planetry system actually forming.

Other investigations by Peter Boyce and William Sinton at Lowell Observatory, using an ingenious interferometric technique that permits of high sensitivity, immediate computer programming, and the rejection of sky (atmospheric) effects, have provided some interesting comparisons between giant and supergiant red stars. Bands due to carbon monoxide have been observed in a giant like Aldebaran (α Tauri), and these bands appear also in supergiants like Rasalgethi (α Herculis), and Mira (o Ceti), but the supergiants also appear to possess strong absorption at wavelengths characteristic of water vapour, so strong that Boyce has referred to such stars as being 'dripping wet'. These investigators have also found an unexpected result for carbon stars, which should in theory display strong carbon monoxide absorption, but fail to do so. They conjecture that since seven stars of this kind which they have examined behave in this way, it may be that they are deficient in oxygen. Should this explanation be correct, then some revision in present ideas of the constitution and behaviour of such stars will be necessary. Infrared studies can also provide clues to the birth of stars, as a recent investigation of the nebula in Orion has shown, but it will be more convenient to postpone discussion of this until the next chapter. It will be clear, however, that the examination of the extreme ends of the electromagnetic spectrum visible from Earth are providing new evidence that is likely to cause considerable re-assessment in our ideas of the universe. Nevertheless, the general picture that has been formulated over the last half century is essentially based on the visible wavelengths; in order to view other new and sometimes startling results in perspective, we must now examine this visual picture.

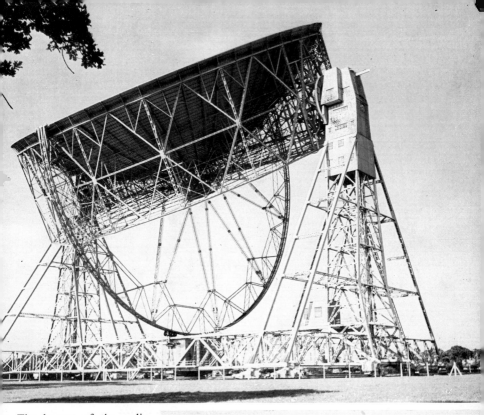

A. The largest of the radio
[t]escopes at Jodrell Bank,
[Ch]eshire. The parabolically
[sh]aped bowl (the dark area seen
[thr]ough the steel framework)
[ha]s a diameter of 250 feet (75
[me]tres) and this is still the largest
[ful]ly steerable radio telescope in
the world.

B. The movable antenna of
[th]e world's first large aperture
[sy]nthesis radio telescope at the
[M]ullard Radio Astronomy
[La]boratory at Cambridge Uni-
[ver]sity. The long concrete pier
[car]ries a railroad track running
[th]e north and south, and ob-
[ser]vations are made with the
[aer]ial at various points along the
[tra]ck.

IVc. The huge 1,000 foot (300 metre) radio telescope at Arecibo in Puerto Rico. This has been constructed in a natural hollow. The bowl reflector lies below an antenna mounted on wire supports at its focus. The bowl cannot be tilted but the small antenna can be moved to various positions which is equivalent to tilting part of the bowl. It is operated by Cornell University.

IVd. Recording signals from a radio telescope. There is, of course, no picture as such seen by a radio telescope, but the signals received are automatically recorded on punched tape and on long paper charts. The triple pen recorder shown here is operated by the 250-foot radio telescope at Jodrell Bank (plate IVA).

CHAPTER 4

The Visual Universe

The most obvious objects in the visual universe are the Sun and the Moon, but away from the glare of urban lights, the stars too make an immediate claim on the attention. Among the stars are some apparently star-like that wander about, for ever changing their positions and always shining with a steady light. These are the planets which, as mentioned in the first chapter, were objects that were the mainstay of pre-telescopic astronomy. In sketching a picture of the visual universe it will be convenient to consider the planets and the Moon first, and then pass to the Sun, stars and other celestial objects.

The only planets for which there is *firm* evidence are those that orbit round the Sun. Whether other stars possess planetary systems of their own is unknown, although it is generally accepted as likely, at least in stars of spectral types from F5 to K which account for about ten per cent of the stars in our own Galaxy. What any other planets may be like depends upon the precise method of their formation, but it is unlikely that they would present wildly different features from the members of the solar system; indeed it may well be that some support life of one kind or another. The solar system consists of nine principal planets and thousands of small planetary bodies, sometimes referred to as asteroids since they present a star-like appearance in a telescope, or, more descriptively, as planetoids.

Many of the planets possess natural satellites and, in general, these are all small compared with the planet about which they orbit. The one exception is the Moon, which is rather more than a quarter the size of the Earth, whereas for no other planet does its largest satellite exceed one twelfth of its primary in size. This has led some astronomers to suggest that the Earth and Moon should really be considered a double planet in which the components are orbiting around one another (as well, of course, as orbiting round the Sun). In this case, the Moon and the Earth must be considered as having formed separately but probably close to one another in space. On the other hand, in recent years, interest has been revived in a theory of the Moon's formation first proposed in the later nineteenth century by George Darwin, son of the famous

49

biologist. Darwin suggested that the Moon and the Earth had originally been one body, but that the Moon had broken away owing to the 'proto' body's fast rotation, which he believed to have been quicker in the past than now. This theory has undergone many vicissitudes, finding favour in one generation only to be rejected in the next, and is at the moment gaining some ground since the examination of the lunar surface by the radio-controlled equipment of the space probe Surveyor 5. This has made an analysis of two small patches of the lunar 'ground' and found that the rock appears to be much like terrestrial basalt, containing similar quantities of the component minerals. If this sample is typical of the whole lunar surface, then it seems that the most abundant elements there are the same as those composing terrestrial rocks – or at least this is the opinion of Anthony Turkevich who directed the experimental analysis – and this would offer some strong, although not conclusive, support for Darwin's hypothesis.

The moon's surface is for the most part pitted with numerous craters, although the side permanently turned towards us also displays a number of large flat plains, known for historical reasons, as 'maria' (seas). The cause of the craters is still undecided, some selenographers feeling almost certain that the majority are the results of volcanic activity in the past, others being as strongly convinced that they were caused by meteoric bombardment. It seems impossible to reach any firm decision about this question at the moment, and it may well be necessary to wait until astronauts visit the Moon and return with rock samples before there is sufficient evidence to solve an intriguing problem: a problem that seems to have a wider significance, since in 1965 the space probe Mariner IV radioed television pictures of the surface of Mars and showed that large craters exist there also.

But whether vulcanism or meteoric bombardment caused the lunar and Martian craters, it is certain that meteors are still in evidence within the solar system. These lumps of rock and metal travel, for the most part, in swarms having very eccentric elliptical orbits, far more elongated than any planetary orbit. Investigations of these swarms has left no doubt about their being intimately connected with comets, which have orbits similar in shape and often identical. Since comets are visible only when close to the Sun, they long remained a mystery, but now it is clear that they are essentially a collection of lumps of meteoric material in orbit round the Sun. This material is heated, and a little of it dispersed each time the comet approaches close to the Sun, and its tail is caused by reactions arising from solar radiation. The dispersal of cometary material occurs along the orbital path and it is this which forms the meteor swarms.

The majority of comets, if not all, are members of the solar system, and form a kind of orbital shell around it, the outermost parts of which extend some thirteen to fourteen billion miles into space: a distance forty times greater than that of the outermost planet. Yet very few comets orbit out as far as this and, generally speaking, the solar system is very much isolated in space, the next nearest star to the Sun lying at a distance of almost twenty-six thousand billion miles. Other stars lie at greater distances still, and only fifteen are known to be within sixty-nine thousand billion miles (i.e., $11\frac{1}{2}$ light years). These, as well as all other stars, are in motion and it is through an analysis of these motions that the visual astronomer has been able to gain an idea of the distribution and behaviour of the stars, and to appreciate the fact that all are members of our own Galaxy – indeed it is such observations that have made it certain that our own Galaxy exists.

The analysis of stellar motions has been an exercise in the application of statistical methods, and has shown that stars possess a preferential motion in particular directions. This was difficult to detect because there is one class of star with a high velocity, and another with a slower velocity similar to that of the Sun, and these are likely to be confused if one is unaware of the existence of the two classes; each star also possesses an individual motion and, in addition, there are corrections to be made for the motion of the Sun, and the orbital motion of the Earth as well as movements of its axis. W. H. S. Monck of Dublin as far back as 1892 noticed that stars of one spectral type seemed to have larger motions than others, and three years later H. Kobold found that, when an analysis of stellar movements was made, there appeared to be a preferential motion along the plane of the Milky Way. However, it was not until 1904 that real order was brought out of the host of observations available, for then the Dutch astronomer J. C. Kapteyn was able to announce that a statistical study of stellar movements showed that he could detect two preferential streams in the motions. The stars in each stream were, he thought, moving with similar velocities, but their directions lay in different parts of the celestial sphere. As they moved, the streams intermingled without affecting each other. Kapteyn interpreted his results as indicating that the stars were part of a rotation of the whole Milky Way system – the Galaxy – and he placed the Sun some 2000 light years from the centre. He considered the stars to be distributed in the form of an ellipsoid of revolution,* the longer axis of which coincided with the plane of the Milky Way, and found that the two streams

* If one takes an ellipse, imagines it pivoted about its shorter axis, and then rotated about this axis, the space traced out is an ellipsoid of revolution.

appeared to rotate in opposite directions. Karl Schwarzschild and Sir Arthur Eddington also made independent analysis, but the full explanation did not come for some years. It was Bertil Lindblad who, in the late 1920s, made new statistical analyses and showed that the two opposite directions of motion obtained by Kapteyn could be resolved if the stars were assumed to possess elliptical orbits in one direction only, the two star-stream hypothesis being a consequence of interpreting apparent motions near the Sun as real rather than relative. J. H. Oort at Leyden developed this argument further with the result that he suggested that the Galaxy was in rotation about a point situated somewhere in the region in the constellation of Sagittarius, which lies across the Milky Way and where the densest star clouds are situated. Oort worked out his suggestions in detail and concluded that the Sun needed 140 million years to complete one orbit of the Galaxy, that the stars closer to the centre orbited more rapidly than those further away from it and, finally, that the Sun lay not 2000 light years but some 19,500 light years from the centre. Lindblad later extended this distance to 30,600 light years and 200 million years for its orbital period, a figure that was also reached by Harlow Shapley at Mount Wilson and later confirmed by other observations.

All these computations were hampered by the dust, gas and density of stars close to the centre of the Galaxy, making it impossible to observe objects lying there. Optically there is little more that can be done, except to analyse further the motions of stars comparatively close to the Sun and of the globular clusters which spread above and below the plane of the Galaxy; as will become evident in Chapter 6, other techniques must be employed to deal with this problem.

The nature and behaviour of stars has been studied by optical astronomers utilizing a background of theoretical and atomic physics, analysis of the H-R diagram, and by detailed studies of the closest star to us, the Sun. It will be convenient to discuss the Sun first, and then apply the information derived from it to other stars. The points that arise from theoretical physics will be mentioned as and when they become relevant, but atomic aspects will be examined in the next chapter in preparation for a discussion of investigations in the more extreme invisible wavelengths of the spectrum.

The Sun, although a G2 dwarf, is large by terrestrial standards, having a diameter of some 865,400 miles and bulky enough to contain 1,303,800 bodies the size of the Earth; its mass is almost 333,000 times that of the Earth. The only body in the solar system to emit light continually, it radiates at a prodigious rate, giving out an energy equivalent to 70,000 horsepower from each square yard of its surface, which has a total of $7\cdot29\times10^{18}$ (7·29 billion billion) square yards. Its surface that one observes through a telescope,* sees visually through mist or fog, or which is photographed and from which most of the radiation is emitted is known as the 'photosphere' and it is on this that dark markings appear from time to time. These are sunspots, in general too small to be visible to the unaided eye, and although they appear dark are, in fact, only so by contrast with the bright photosphere. Sunspots grow in size once they have appeared and usually occur in groups, each spot being 20,000 to 30,000 miles in diameter, although exceptionally large spots with lengths of up to 90,000 miles and widths of 60,000 have been observed. They appear only north or south of the solar equator – never on it – and seldom reach latitudes greater than 45°, while there is a cycle of maximum and minimum numbers that has a period of eleven years between one maximum (or minimum) and the next. The temperature in a sunspot is some thousand degrees less than the surrounding photosphere, and spectroscopic studies show evidence of what is called the Zeeman effect: the spots are magnetic and when they occur in pairs, the members possess opposite magnetic polarities. What is more, if the leading† member of a pair has a particular polarity, the same polarity is followed by all subsequent leading spots in that hemisphere for the rest of the sunspot cycle. At the next cycle, leading spots in that hemisphere present the opposite polarity. Sunspots may be envisaged as tornadoes of electrified gases, although most of the rotation occurs below the photosphere. Their magnetic polarity is thought to be due to the interaction between the general magnetic field that the Sun possesses, and the magnetic field caused by the motion of the electrified gas, since the movement of such a gas is similar to an electric current which invariably has a magnetic field associated with it. The opposite polarity of a pair of sunspots is then explained as due to the tornado (one spot) running under the photosphere and emerging again to give the second spot: in this case the gases will be rotating in opposite directions at each spot and this would give rise to an opposite polarity. Why there is a maximum and minimum every eleven years is still not satisfactorily understood, although it is the general consensus of opinion that it must be due to electrical and magnetic effects below the surface of the photosphere. Horace Babcock of the Mount Wilson and Palomar Observatories has, however, suggested

* In practice the Sun is too bright to observe directly through a telescope: it would blind an observer and indirect methods, such as projecting the solar image, must be used.
† The leading spot of a pair is the one which lies ahead in the direction of rotation (W to E) of the Sun.

that since the Sun rotates unevenly, the higher latitudes rotating more slowly than those close to the equator, lines of magnetic force travelling from the Sun's north magnetic pole to its south magnetic pole will become distorted where they pass under the photosphere because there the gases are subject to a lagging effect. This lagging behind of the gases would be sufficient, he believes, to distort the magnetic field so that the lines of magnetic force are drawn out into lines: under the photosphere they will become identified with the flow of ionized gases, and the concentration of magnetic energy, thereby formed, will burst out from time to time to cause spots.

Above the photosphere lies a sphere of thinner gas that, when visible at the end of a total solar eclipse just before the full glare of the photosphere returns, has a reddish colour and is called the 'chromosphere'. It merges with the photospheric gases at its lower levels and extends above it some 5000 miles. Its chemical composition is uneven: iron and other metals in an ionized state extend up to a height of some 1500 miles, hydrogen about 7500 miles, and ionized calcium higher still, while ionized helium and oxygen are also present. Through the chromosphere there extend prominences – flame-like structures of hot wispy gas associated either with sunspots or appearing over latitudes of around 45°. These range in length from a few hundred miles to hundreds of thousands of miles, and they are an evanescent phenomenon, a prominence often lying along the Sun's lines of magnetic force, with its gases in continual motion; the velocities of these gases can range from as low as a few miles per second to hundreds of miles per second. Prominences seem to be composed mainly of hydrogen, helium and calcium, but in those that are more eruptive, normal and electrified metals are also to be found.

Above the chromosphere extends the corona, which may be seen as a pearly coloured light during the few minutes of a total solar eclipse. The shape of the corona varies, at times of sunspot maximum appearing as a spherical envelope around the Sun, but becoming elongated in the direction of the solar equator at sunspot minimum. It is in reality a thin atmosphere extending outwards at least some 375,000 miles beyond the photosphere, and spectroscopic evidence makes it clear that it becomes hotter at increasing distance, although temperature in this case is measured by motion of the ionized atoms of the gas, and, although real, is rather different from the ordinary temperatures of hot solid bodies and hot dense gases like flames with which we are familiar.

As a whole the Sun is opaque although gaseous, and this is the reason that other bodies cannot be observed through it. Indeed photographs of the solar disk show a darkening at the edges; this is caused by the opacity of the outermost layers of the photosphere, since one is there observing through a greater thickness than when viewing the central parts of the disk. The corona, however, is so tenuous as to be transparent and it offers very little obstruction to the passage of radiation from the photosphere into space.

Even though astronomers cannot see through the Sun, there is nothing to prevent them from dealing with its structure by theoretical methods, and this has been the course adapted for many years with singular success. The main problem to be solved is to find how the Sun has maintained a steady – or reasonably steady – radiation at a high rate for a period of not less than three billion years (a figure obtained from geological and mineralogical considerations about the age of the Earth and, hence, of the solar system). There is no known chemical reaction such as burning which can generate radiation at the rate required (70,000 horsepower per square yard) from the Sun's material for so long a time, nor is the energy released by the contraction of the Sun under its gravitational pull sufficient, and a process must be sought elsewhere. It was Eddington who first suggested that some kind of process whereby atoms were broken down inside the central parts of a star was the only likely possibility, and in 1932, Robert Atkinson, another English astronomer, worked out a theoretical model of the way in which matter could be annihilated, and calculated the amount of energy that would be generated from such annihilation. To prosecute this line of research further and in greater detail took time. In 1939 Hans Bethe and C. F. von Weizsäcker worked out cycles of nuclear activity more extensive than Atkinson's, but only after the practical and theoretical experience of the use of atomic weapons during the Second World War did these ideas assume more than a theoretical interest. Now the energy process in the Sun is far better understood, and seems to be one of what has come to be called nuclear fusion – the kind of process that occurs in a hydrogen bomb. Here the central core or nucleus of an atom fuses with the nuclei of other atoms, forming a new and heavier atom and, at the same time, releasing a vast amount of energy. Calculation shows that this is more than sufficient to keep a star like the Sun radiating for billions of years.

The nuclear fusion process occurs in the central parts of the Sun and of other stars, and might be expected to explode the star were it not for the gravitational pull on the gaseous material lying above it. This presses the gases inwards towards the centre of the star and the heat and other radiative energy exert a force as they move outwards to escape. Towards the centre of a star, the difference in temperature between the central core where nuclear fusion is occurring and its surrounding gases is so great that heat is transferred by convection. In other words, the very hot gases move upwards to slightly cooler regions and the cooler gases move in to take their place, are heated in their turn, move upwards, and so on. But this only occurs where there is a large difference in temperature between one part of the star and another, otherwise radiation passes from the inner parts to the outer by 'radiative transfer' – by the radiation itself moving directly outwards. This radiation may be thought of as being propelled by its pressure on the gas particles inside the star, for although in the everyday world with which we are familiar the pressure of radiation on objects is not observable, this is because the objects are large compared with particles of gas, and radiation is of a limited intensity. But where one has a source of great intensity like a star and particles are as minute as gas particles so that the gravitational pull on each particle is small, radiation pressure becomes an important factor.* But few stars are as simple as this and more elaborate 'models' of stellar interiors have been prepared. Before discussing them, it is necessary to look again at the H-R diagram (figure 9).

The H-R diagram shows that the main corpus of stars – the main sequence – goes from hot intrinsically bright objects to cool dimmer stars (the red dwarfs), and in his early analysis of this, Henry Norris Russell suggested that the sequence showed that stars began as hot and bright and then, after radiating away their energy, ended up as cool and dim. This was an obvious enough supposition, running parallel to the behaviour of a heated black body, but after the realization that stellar energy is derived from nuclear fusion processes, this hypothesis had to be discarded. Stars do age and evolve in the sense that they move from one spectral class to another, but not in the way previously supposed. However, the terms 'early' and 'late' are still often used for referring to the positions of stars in the H-R diagram, and taken in Russell's original sense, so that O, B and A types are early and K, M and other red stars are late.

The new evolutionary ideas arise from the realization that nuclear fusion takes place in the centres of stars, that they may be expected to increase in heat for part of their lives, rather than cool from birth to death. The fusion that occurs may take a number of forms, the precise heating or thermonuclear reaction depending on the temperature of the centre core of the star and also on the presence or absence of some heavier atoms, since all stars appear to be composed primarily of hydrogen. In general, there are two main kinds of thermonuclear

---

* Radiation pressure is, in fact, proportional to the fourth power of the temperature (*i.e.*, to T^4).

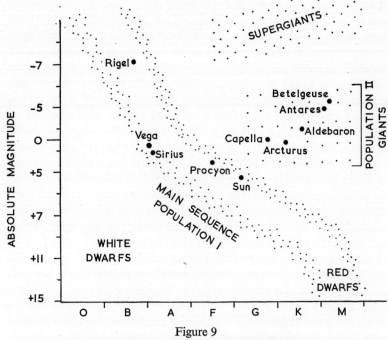

Figure 9

The Hertzsprung-Russell (H-R) diagram. Here the different classes of stars are plotted against a measure of their intrinsic brightness (absolute magnitude). Some of the better known bright stars are named. The terms 'population I' and 'population II' are described on page 58.

reaction that are thought to be occurring in most stars: one in which the nuclei of hydrogen atoms combine, and a second where carbon oxygen and nitrogen nuclei are involved as well as hydrogen. Since hydrogen nuclei are composed solely of the nuclear particles known as protons, the first is known as the proton-proton reaction, and the second as the carbon cycle. The carbon cycle needs a very high temperature – something more than ten million degrees K – before it can occur, but the proton-proton reaction can take place at a much lower temperatue. In the proton-proton reaction the generation of energy varies with the temperature, becoming greater the higher the temperature. In addition there are other thermonuclear processes that will happen at even higher temperatures, but these are not the rule for the majority of stars.

With this in mind, the general consensus of opinion is that a star forms by condensing out of an amorphous mass of nebular gas; this can occur either in single or multiple condensations (in the latter case a cluster of stars is formed), but whichever is the case, the condensation begins due to the irregular distribution of gas in a nebula. The stars form from the denser parts owing to gravitational attraction. At the start there is a small quantity of material, but as the body grows so it attracts more material, until the gas in the centre becomes subjected to great pressure and, because of the gravitational energy released as the material falls inwards, to an increasing temperature. At some point the internal temperature will reach a value that is sufficient for the proton-proton reaction to occur. This builds the temperature up further and some of the radiation escapes – the star begins to glow. The process now continues until an equilibrium is reached when radiation pressure and the pressure of hot gases outwards is balanced by the gravitational pull inwards of all the gaseous material. The temperature of the central core is some million degrees K. After this the star continues to radiate for a long time in a balanced way, neither contracting nor expanding to any great degree, and lies in the main-sequence of the H-R diagram.

Not all stars enter the H-R diagram at the same point, some enter at the late end (M), others towards the earlier end at A or F. It is the massiveness of the newly formed star that determines this, since the more material there is, the greater the gravitational energy released and the hotter the internal temperature – in consequence the more massive stars have higher internal temperatures and so the proton-proton reaction is faster, more energy is generated, and the star shines more brightly. It enters at the earlier end of the main sequence. In general two types of star have been recognized, those in the area of the Sun and those which are to be found in globular clusters. In 1944 Walter Baade of Mount Wilson and Palomar Observatories drew attention to the fact that these differences were to be noted in other parts of our Galaxy, and in other galaxies as well. Baade classified these stars as of two different 'populations', population I referring to the stars in the neighbourhood of the Sun and the first to be recognized, while population II contained the globular cluster stars. Population I frequents the regions of the galaxy where gas and dust are abundant, and its brightest stars are blue, lying on the main sequence, while its red giants are not exceptionally bright. In the case of population II stars, these occur in regions where there is little dust and gas, with red giants intrinsically brighter than those of population I. The development of the stars in each population is different, those of population I seeming to contain more heavy chemical elements in their composition than those of population II; we shall now briefly trace the development of each type separately, beginning with population II for reasons which will become obvious in a moment.

Population II stars, composed for the most part of hydrogen, condense in the way described, but their masses are such that they enter the main-sequence at a point around M, and are red dwarfs. As the proton-proton reaction continues, hydrogen is converted into the heavier element helium, and this acts in a way analogous to ash since it will not 'burn up' (in a nuclear reaction) and radiation has difficulty in escaping: the internal temperature of the star rises. In due course this is evident as an increase in intrinsic brightness, and the star moves up the main-sequence to an earlier position. Because of the greater internal temperature, there is an acceleration of the proton-proton reaction, and still more heat is generated. The star moves further up the main-sequence – the distance depending on the way in which the material inside the star is mixed. If the elements in it are distributed homogeneously, then the increase in temperature will continue until the star moves up to the later end of spectral class A. But if, as is more likely, the distribution inside it is inhomogeneous, then it will leave the main-sequence before A is reached. The homogeneity or inhomogeneity appears to be due to the rotational velocity with which the star begins its life the faster the rotation the better the mixing, and the more homogeneous it will be.

If, while the star is on the main-sequence, a considerable amount of heavy elements is built up during the proton-proton reaction (which will occur if the temperature rises enough so that the reaction begins to occur more speedily), the carbon cycle can come into play. The star will then rise to an earlier position along the main-sequence before leaving it. At this stage, the star will have already developed a central core where there is mainly helium ash, around which lies a shell where hydrogen is consumed in thermonuclear reactions. Above this is a large volume of gas in the outer layers of which heat is conducted by convection rather than by radiation. This is probably the form of the Sun, although it is not considered yet to be close to the point where it will leave the main-sequence.

When such a star does leave the main-sequence, it moves upwards to the right of the H-R diagram and becomes a red giant. The reason for this is that as the hydrogen shell 'burns', the helium core enlarges, and the hydrogen shell where nuclear reactions are occurring moves outwards; the star expands. As it does so, the outer gases become cooler and the star assumes giant proportions. After this, further changes occur inside, and more heat is conducted away by convection: in consequence the expansion ceases, more hydrogen is consumed, and the helium core increases to such a degree that it begins to contract under its own weight. This contraction heats the star still further and the expansion of the outer gases moves it into the supergiant position on the H-R diagram. What now happens is probably that the star begins to contract, and that as the internal conditions undergo changes, so does the star's luminosity, and it becomes a variable, perhaps of the RR Lyrae type. However, a star is not thought to remain very long in this stage – perhaps some hundreds of thousands of years only – but rather to contract further with an increasingly dense core, so that it finally becomes a dim, small, dense star radiating strongly in bluish wavelengths but known as a white dwarf. The diameters of such stars have been calculated to be about 22,000 miles (one fortieth the diameter of the Sun) at the most, but some are far smaller; although they are extraordinarily dense – a lump of material about the size of a matchbox would weigh something over twenty tons – their total mass is small. On the way to becoming white dwarfs, stars will have to lose a proportion of their material and it may well be that some, at least, pass through the stage of being novae. What happens after the white dwarf stage is uncertain, but there is probably further contraction that will turn the body into a neutron star or even an hyperon star, types that it will be more convenient to discuss in the next chapter.

The development and life cycles of population I stars are somewhat different in detail, since these stars contain a higher proportion of heavy elements than those of population II. During the life of a population II star, heavier elements are synthesized from the lighter ones with which the star begins by the thermonuclear processes that occur in the star's core. Some of this heavier material – perhaps a considerable proportion if the star passes through the nova stage – is ejected during the star's lifetime and will take its place among the dust and gas that form the nebulae, and from which there is good reason to believe that new stars are being born. It appears, then, that population I stars must be younger than those of population II, and they are sometimes referred to as second or third generation stars.

Condensing, like population II, from an amorphous mass of gas, the population I type star heats up owing to gravitational contraction. However, as soon as the proton-proton reaction becomes operative, the star's core is increased in mass by the helium ash, but to a greater extent than in population II stars since more heavy elements are present. Further contraction occurs and the temperature rises still further. The presence of heavy elements provides fuel for the carbon cycle to take over as soon as the temperature is sufficient, and as this form of nuclear fusion creates more energy than the proton-proton reaction, the star is bright. It enters the main-sequence in the earlier regions and moves up to spectral classes

B and O. In due course these stars, like those of population II, leave the main-sequence, become giants, supergiants, variables and then dwarfs, but there are detailed differences. The population II star synthesizes some but not all heavy elements, so the stars of population I start with an advantage since they already possess some of these elements; because of this, other thermonuclear processes besides the carbon cycle can occur and more complex heavy elements can be synthesized in them than is possible in population II.

It must be emphasized that this whole description of stellar development is in many ways a tentative one. Some stages – the earlier ones – have been computed in detail, but others are so complex that there is

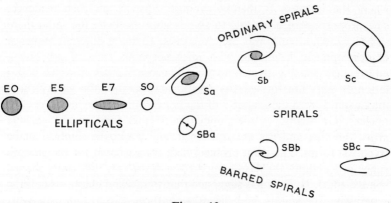

Figure 10

Classification of galaxies by shapes. This classification made in the 1930s by Edwin Hubble possibly indicates galactic evolution if we read it from right to left.

still much computation to be done. Again there is the assumption that stars begin with a composition mostly of hydrogen – in the population II class almost entirely so – although this assumption appears valid from a consideration of other information. Yet there are many questions still to be answered and, in particular, it is becoming clear that stellar development is also linked with the development of galaxies, though here we are in an even more tentative realm. The first attempt to produce a galaxy development hypothesis was made in 1928 by James Jeans who, using the observations of Edwin Hubble of Mount Wilson and his classification of galaxies by their shapes (figure 10), suggested that galaxies were formed from hydrogen gas as huge spherical objects in slow rotation. Gravitational attraction of this giant globule of gas would cause contraction, the speed of rotation would increase and the galaxy would become progessively flattened. Meanwhile, the population I stars were formed, lived their lives and ejected their heavy elements into space, forming dust that was thrown out by rotation into the spiral arms that had now formed and where, indeed, population I stars are to be seen. In other words spiral galaxies are far younger than ellipticals.

Hubble himself conceived of a different evolutionary picture. He thought that galaxies began as type E0 and then became more flattened, moving to E7 and thence to spiral form. If Hubble's idea were correct, then in the earlier stages of the universe there should be more elliptical galaxies than any others, but observations made in the most distant parts of space from whence light has taken longer to reach us, and where the universe is thereby younger, display no such evidence. The proportion of ellipticals to spirals appears to be the same there as nearby. An attempt was therefore made to find a different sequence of development, beginning with an amorphous mass of gas giving an irregular galaxy with a comparatively dense nucleus, in which population II stars are formed. As rotation accelerates, the galaxy turns into a spiral, with the majority of the population II stars in the central nucleus. Gradually, as the population I stars form in the galactic arms, the dust and gas are consumed, and a smooth elliptical shape (type E7) forms. The main problem here is to account for the smooth shape of an elliptical galaxy compared with the very differently shaped spirals, and J. H. Oort has suggested that spirals and ellipticals develop separately. The spirals, he believes, were formed from thin expanding gas at those points where there were local increases in density. Eddies and collisions of gas particles would then give rise to heating and the formation of denser clouds, while the angular motion that was originally present would result in a rotation of the galaxy and the birth of a number of stars. So far this appears similar to the Hubble hypothesis, but Oort points out that there may well have been patches of original gas – protogalaxies – that possessed little angular motion, and others that had a good deal; the former would turn into ellipticals and the latter into spirals. Spirals, because of their comparatively high speeds of rotation, would be formed since their fast rotating protogalaxies would contract perpendicularly and spread out into disks, with the gas forming them thrown outwards and into spiral arms. Protogalaxies with slow rotations would retain a spherical or elliptical shape and the gas, not being whirled about so energetically, would all condense into stars. On this hypothesis, population II stars and globular clusters are formed in the protogalaxy stage. At the moment the situation is unsettled but, as will become evident in later chapters, this idea seems to be rather more in accord with the most recent observations made in invisible wavelengths.

What is no longer in doubt is that there are local inhomogeneities in the distribution of galaxies in space, although in general they appear to be uniformly spread. The local inhomogeneities are known as clusters of galaxies (see plate IIIa), and here and there, within a cluster, a double or triple galaxy is to be found. Our own Galaxy with the two Magellanic Clouds is one such triple system and the spirals (plate IIIb) form a double system. Fritz Zwicky of Mount Palomar has investigated the spiral (M 51) with his photographic techniques of superimposing pictures taken with different colour filters, and has been led to the surprising conclusion that while the yellow-green stars form a barred spiral,* the blue population I stars give the appearance of being members of an ordinary rather open-armed spiral. These separate galaxies – if that is what they are – seem to co-exist in close proximity. Certain pairs of galaxies have also been observed with bridges of what seem to be gaseous material stretching between them, and whatever the answer to the birth and development of galaxies may be, it is becoming clear that they can no longer be considered as objects that undergo very slow development, and it may well be that they will soon be fitted into some very large scale correlation of the universe.

* Galaxies have been classified according to their apparent shape. Those with no visible spiral structure are called 'elliptical' since most of such galaxies display various degrees of oblateness. The spirals on the other hand are divided into two main classes – those which are straightforward spirals and those in which the central region is long and dense: the former are spirals and the latter barred-spirals. In both spiral classes, the outer parts of the spiral may spread out to give open arms, or be closed so that the outline is almost circular.

CHAPTER 5

Physics and the Extension of the Spectrum

The general behaviour of radiation as evinced in stellar spectra has already been mentioned, but it is now necessary to pursue this matter further, since the generation of all electromagnetic frequencies – visible and invisible – must be considered. Such an investigation must begin with a brief description of the atom and the particles that compose it, although a word of caution should be offered about the use of the word 'particle'. Our everyday experience is based on what we sense directly of the world around us and our laws of the behaviour of bodies which are immensely vast by sub-atomic standards. We experience, in fact, a macroscopic world where the forces between bodies are primarily either gravitational or magnetic. A body is something solid, and when the word particle is mentioned it is natural to think of something like a very small billiard ball or pea. Yet sub-atomic particles behave like waves in some experiments and particles in others, and it is hard to conceive of them in this way: one must therefore appreciate that while particle is a useful connotation, it must never be taken too literally; if it is, then some of the explanations are bound to produce unnecessary difficulties. What is more, it must be emphasized that while the whole of atomic and sub-atomic or nuclear physics is based on an hypothesis of particles, no one has even observed any of these particles as such. What has been observed is a host of phenomena and interactions for which the most convenient and satisfactory explanation is in terms of particles. Sometimes this leads to the supposition that there are transitory particles – the mesons and hyperons – that have a lifetime of a couple of millionths of a second to ten billion times less but nevertheless it is better to use these than to try to explain empirical evidence by the existence of new forces of a peculiar kind.

The simplest atom is that of hydrogen, and it is usually described as having a nucleus around which orbits one electron. An electron is a sub-atomic particle which possesses a minute negative electric charge, and a very small mass (about 9×10^{-28} grammes). It may be as well to

Vₐ and B. At A (*top*) the lunar crater Tycho photographed using a large optical telescope on earth. At B (*below*) the same feature built up photographically from radar echoes transmitted and received on earth. The definition of the radar picture is about as good as that of the optical photograph but considerable improvements may be expected. The 'surface' given by the two methods is slightly different since radar penetrates a short distance below the surface.

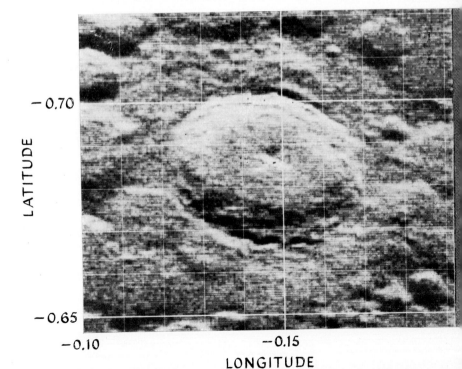

LATITUDE

−0.70

−0.65

−0.10 −0.15

LONGITUDE

VIA. Spiral galaxy (NGC 3031) in Ursa Major. The photograph, taken with the 200-inch telescope, clearly shows the two symmetrical spiral arms and the bright and dark absorbing matter in the nucleus. Some individual stars can also be seen. This galaxy is similar to our own.

point out here that negative is the name given to an electric charge similar in kind to that obtained when a piece of amber is rubbed with a piece of fur, an observation known to the Greek philosophers, whose word for amber was, in fact, 'ēlektron' – while a positive electric charge is that which arises when a substance such as glass is rubbed: the words are arbitrary, and a legacy of historical usage, but they do indicate that the charges are different – positive attracting negative. All that is important is that they be applied consistently. The mass given for the electron is that which it would have when stationary, since experiment shows, and relativity theory states, that what is termed the mass increases when the electron – or any other body, large or small – moves at a high velocity.

The nucleus of the hydrogen atom consists of one particle – a proton. Its 'rest mass' is a little more than 1836 times that of the electron, and this means that the main mass of this atom – and indeed all other atoms – is centred in the nucleus. The proton has a positive electric charge equal in intensity to the negative charge of the electron; with the atom in its normal state having both electron and proton present, the whole is electrically neutral since negative balances positive. The electron orbits the nucleus (proton) and there are a number of possible but quite definite orbits it can take – the one nearest to the nucleus being known as the ground state. When the atom receives energy (by heating, from a beam of light, etc.) it jumps from an inner orbit to one further out. Here it remains for a short time, and then falls back to the orbit it originally occupied; as it does so, it emits energy. The energy required for the electron to move from one orbit to another is a quite specific amount – a different quantity for any two orbits, but nonetheless precisely calculable. The frequency (and hence the wavelength) that is absorbed when the electron jumps outwards – or emitted when it returns inwards – depends upon the different orbits involved and nothing else. Photons (energy particles) with these frequencies are absorbed or emitted, one photon of a particular wavelength for each jump.

In examining the emission spectrum of glowing hydrogen gas, a series of lines was discovered whose positions (wavelengths) could be specified by a mathematically computed series of numbers. This was found in 1885 by Johann Balmer, while additional series of lines were found soon after by F. Paschen in the infra-red and by Theodore Lyman in the ultra-violet wavelength ranges, but it was not until the advent of a theory of the atom and specific orbits that a theoretical explanation could be given. This was primarily due to the work of Niels Bohr in 1913, and the ways in which some of the different wavelengths of hydrogen lines

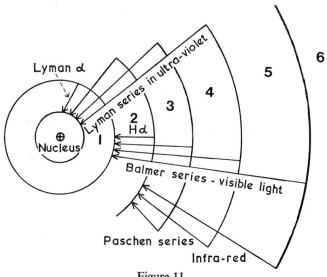

Figure 11

A 'plan' of the different energy states of the single electron orbiting the nucleus
of the hydrogen atom. The 'series' referred to are those series of lines observed
in the laboratory.

are due to the photons emitted (or absorbed) during orbit jumps of the
electron, are indicated in figure 11. It will be evident that the ultra-violet
wavelengths are emitted when the electron is performing jumps to and
from the ground state, jumps in which the photons have the greatest
energies since the greatest energies are involved in moving to and from
ground state. The reader should notice particularly the Lyman alpha
(α) line, generated by jumps between the next outer orbit and ground
state, since we shall have particular occasion to refer to this line – the
longest wavelength hydrogen line in the ultra-violet – in a later chapter.
The visual wavelength or Balmer series of lines is due to jumps between
the orbit next to the ground state and other orbits, and involves energies
less than those of the ultra-violet series, but greater than the energies
required for the absorption or emission of the Paschen infra-red
series.

It has been found helpful to specify the energies involved, and the
units used are 'electron volts' (abbreviated ev). The electron volt may
be defined as that electrical charge which moves an electron from one
place to another with a certain force: more specifically, if an electron is

placed between two metal plates in a vacuum, and the potential difference between the negatively charged metal plate and the positively charged one is one volt (by connecting the plates to a one volt electric battery), then the electron (being negatively charged) will be repelled from the negative plate and attracted towards the positive one. By the time it reaches the positive plate it will possess a certain energy due to its motion – kinetic energy – and this energy is one electron volt.* Each photon can be said to possess an energy of so many electron volts, and thus the jumps will be due to specific energies or a specific amount of electron volts (or fractions of them). For example, the first line of Balmer's series in the visual range of the hydrogen spectrum, known as the H alpha (Hα) line, has a wavelength of 6563 Å, and is due to the electron jumping between orbit three and orbit two (orbit one is ground state); it requires an energy of 1·89 ev. The longer wavelength lines in the spectrum require smaller, and the shorter wavelength lines greater, energies. To take further examples, for the Paschen α line in the infrared with a wavelength of the order of 18,751 Å, the electron moves from orbit four to orbit three and the energy involved is no more than 0·66 ev, but when we come to the Lyman α line at a wavelength of 1216 Å in the ultra-violet, an energy of 10·2 ev is required to make the electron move from its ground state to orbit two.

To cause the electron to jump from ground state to orbit six required an energy of 13·2 ev, but, if an energy greater than this is applied, the electron will be made to jump further than the sixth orbit and it will then become detached from the nucleus. An energy of 13·54 ev or more will achieve this, and the hydrogen atom, bereft of its solitary electron, will consist of a proton only: it will therefore possess a positive electric charge. An atom in this state is said to be ionized.

Hydrogen is the simplest of atoms, with its one orbiting electron. The next atom in complexity is that of helium, where there are two orbiting electrons and hence a double negative electric charge outside the nucleus. To compensate for this and so keep the atom electrically neutral, the nucleus must possess a double positive electric charge; in helium this means that the nucleus must contain two protons, not one as in the case of hydrogen. The two electrons move in a similar kind of ground state orbit, but there is a factor that prevents these negative particles from

---

* This energy can be expressed in mechanical terms. For instance, if a mass of one gramme is given a push that accelerates it so that its velocity increases by one centimetre each second, the force of the push is said to be one dyne. If the push of one dyne moves through one centimetre to cause this, work is done, and this work is defined as one erg. One electron volt is 1·6 millionths of an erg.

being precisely the same, and this is their spin. The spin can best be thought of as the electron rotating on its axis as it travels in its orbit, although this has not – and cannot – be observed, and the concept was introduced in order to deal in a satisfactory way with atomic behaviour; in particular, with the behaviour as demanded by the theory of relativity. In practice there appear to be certain empirical results that can be best explained if one envisages the spinning of an electrically charged particle, since various magnetic and electrical effects follow according to the basic principles of electricity. The empirical results that can be explained by the hypothesis of electron spin are certain effects on spectral lines when the atoms concerned with the radiation are within a magnetic field. The Zeeman splitting of the lines, referred to in Chapter 3, is more complex than can be explained if the electron is simply moving in an orbit, but this complexity can be accounted for satisfactorily if the electron is supposed to spin. Observation of very fine spectral lines also makes it clear that an electron appears to spin in a simple way, with its axis of spin fixed, and the rotation either clockwise or anti-clockwise: there is no need to invoke the idea of electrons spinning with their axes tilted over in one direction or another.

The fact that helium possesses two electrons means that it can undergo two stages of ionization: singly ionized – when it has lost one electron only – or doubly ionized when it has lost both electrons. The energy required to ionize helium is greater than that for hydrogen: to cause single ionization an energy of 24·48 ev is required, and to cause double ionization, the energy must be at least 54·17 ev. Since the helium nucleus contains two protons, helium atoms are at least twice as heavy* as those of normal hydrogen, and as one moves to more complex atoms still, they become heavier as more and more protons are involved in the nucleus to balance the electric charges of the orbiting electrons.

The two electrons of a helium atom, orbiting at a similar distance from the nucleus, fill what is called the first 'shell'. For reasons that seem well established, it has become an accepted view of atomic construction that there are not only a specific number of orbits in which an electron may move, but also that electrons are collected into shells, all the electrons in one shell having their orbits, as it were, on the surface of that shell. Again, from observations on spectral lines and from theoretical considerations, it has become clear that the number of electrons in any one shell is limited, although the numbers are not the same for each shell. For example, in the helium atom there are two

---

* If the helium nuclei contain neutrons as well, they will be still heavier.

electrons in the shell, and in all other atoms, irrespective of the number of orbiting electrons they may have or the composition of their nucleus, there are never more than two electrons in the innermost shell. For a substance like lithium, which has three orbiting electrons, the third electron is forced to orbit further out than the two inner ones, and it orbits in the second shell. This is so for other atoms like carbon (six electrons), nitrogen (seven electrons), oxygen (eight electrons), up to the gas neon. Neon has ten electrons, two in the inner shell and eight in the outer. But eight is the maximum number of electrons that the second shell can hold, and for the next heavier atom, sodium, with eleven electrons, the outermost must go into yet another shell. And so the process repeats as one passes from one element to the next heavier element, but the concept of shells is of more than theoretical interest. The power of elements to combine chemically depends on their electrons, those with a closed shell outside, such as helium and neon, being chemically inert,* while those with one or only a few electrons in their outermost shell are chemically active. Again, if an atom has electrons in a closed or full shell and none outside – as is the case with helium and neon – then considerable energy is required to cause even single ionization; but if there are, say, only one or two in an incompletely filled shell, ionization requires only a comparatively small amount of energy. The astronomical significance of this is probably obvious – stars where electrified or ionized helium is found must be very hot and bright since there is plenty of energy to provide the comparatively high electron voltage to cause such ionization: ionized hydrogen also requires a high ionization voltage compared with metals like iron, nickel, magnesium, sodium, etc., which have a few electrons in an outer shell – these are less strongly bound to the nucleus and so single ionization is possible at lower energies. Oxygen, nitrogen and carbon may best be thought of as intermediate in the energy they require, and with these facts to hand, the spectrum can provide definite evidence of the energies available.

Only hydrogen atoms have their nuclei composed of protons alone, although the number of protons is always equal to the number of orbiting electrons; the others possess a second kind of nuclear particle or nucleon, known as a neutron which, as its name implies, is electrically neutral. The mass of the neutron is almost the same as that of the proton, so that its presence in the nucleus gives a heavier atom, but one in which

---

* The term inert has been used for many years since until recently such elements could not be combined in the laboratory. In 1962 some compounds containing an 'inert' element were at last produced, but the term is convenient as such elements are inert in most reactions.

the charge remains the same so that no extra orbiting electrons are required. The neutron provides considerable variety in atoms, for although the chemical properties of an atom depend upon the protons in the nucleus and the number and arrangement of the electrons, the absence of a neutron or the presence of additional neutrons gives atoms that are lighter or heavier than normal, but with the same chemical properties. Such atoms with masses different from normal are known as isotopes and are of significance in atomic rather than chemical reactions, as will become evident in a moment.

It was as far back as 1896 that Henri Becquerel discovered that uranium, the heaviest of the natural elements, emitted rays, and did so quite spontaneously. The rays appeared to be similar to the X-rays discovered by Wilhelm Röntgen the year before, since, like X-rays, they could make gas in a vacuum tube conduct electricity. The subject was investigated chemically by Pierre and Marie Curie in the years that followed, and they discovered that some other heavy elements also behaved in a similar way. In particular, they found that the emission of radiation – the radioactivity as they called it – appeared to occur within the atoms of the substance, since it seemed to be unaffected by all the chemical processing necessary for isolating the elements. Further research by Ernest Rutherford, Frederick Soddy and others took matters further and the radiation was discovered to be of three kinds, named for convenience α (alpha), β (beta) and γ (gamma) radiation. The α radiation was found to possess a positive electric charge, the β radiation a negative charge, while the γ radiation showed no electric charge at all, but was merely a very penetrating radiation that we now know to be of shorter wavelength than X-rays, and which has already been mentioned in Chapter 2.

The β radiation turned out to be a stream of electrons with various velocities and, therefore, not really a form of radiation at all, while the α radiation was again found to be a shower of particles, all with the same velocities. These α-particles possessed a double positive electric charge, were four times the mass of the nucleus of the hydrogen atom and proved to be the nuclei of helium atoms, each containing two protons and, in addition, two neutrons.

The phenomenon of radioactivity was clearly seen to be a result of atomic disintegration. Without pursuing the matter in detail, it is worth noting that when radium disintegrates, an α-particle is emitted, with the result that the 88 protons and 138 neutrons that form the radium nucleus are reduced to 86 and 134 respectively: the radium is transmuted into the inert gas radon. Radon decays, also by emitting an α-particle, and becomes polonium, which itself emits an α-particle and is transmuted into an isotope of lead. The isotope also decays by emitting a β-particle – an electron – which means that its mass remains virtually the same but the charge of the nucleus is increased, since the electron emission arises from the decay of a neutron into one electron (which is ejected) and one proton. The new chemical substance thus formed is an isotope of bismuth. It was thus established that chemical elements do become transmuted, although not in the way envisaged centuries earlier by the alchemists.

The decay of a neutron is a normal phenomenon, at least outside the nucleus. Neutrons split into a proton and an electron and they do this fairly rapidly: for instance a collection of neutrons outside an atomic nucleus will disintegrate quickly enough so that only half their number will remain as neutrons after twelve minutes, with half the remainder disintegrating in the next twelve minutes, and so on. The period twelve minutes is known as the half-life of the neutron, but it is true only for neutrons outside the nucleus; within the nucleus there is a regular exchange between neutrons and protons, the neutron becoming a proton when it gives an electron to another proton, which thereby becomes a neutron. This regular exchange of identities between protons and neutrons is believed to be the cause of the very strong force which binds an atomic nucleus together.

In spite of the behaviour of neutrons and the emission of electrons during radioactive disintegration, it is found that electrons cannot exist as such inside an atomic nucleus. When a neutron decays, the electron is, as it were, created there and then, in a way analogous to the creation of a flame when burning commences. The problem here, however, is that in a radioactive element, when a neutron decays and emits an electron which leaves the nucleus instead of combining with another proton, one can argue that a new electric charge is generated, and the physicist is adamant that new electric charges cannot suddenly come into being. Relativity theory is quite happy with the creation of mass from energy since both mass and energy are supposed to be equivalent; but not so electric charge. Physical theory therefore demanded the generation of a positive charge to balance the negative charge of the electron. A particle called a positive electron would achieve this, and in 1928 Paul Dirac had foreshadowed such a particle. In 1932 it was found to be a way of explaining details of nuclear reactions then actually observed in the laboratory. The positron is, however, a very transitory particle, having a life for the most part measured in thousand billionths (10^{-12}) of a second. A positron ends its brief life by combining with an electron, whereupon both are annihilated and transformed into energy (γ-rays).

The disintegration of radioactive materials raises another problem, and this is the variety of energies with which electrons are emitted. The nucleus is bound with the peculiar exchange forces, but in radioactive decay, particle identity changes occur without an exchange in the nucleus; even so electrons ought in theory to be emitted with certain discrete amounts or quanta of energy, but this is not found to be so – their energies have all kinds of values. All the same, the principle of discrete radiation is so fundamental and so admirable an explanation of, for instance, the generation of absorption and emission lines in spectra, that quanta must be retained in the physicist's hypothesis, and some other explanation than its rejection found to account for the multiplicity of electron energies observed. If it is assumed that all the electrons are ejected with energies less than the maximum quanta they could have – and there is good reason to suppose that this is true – then the problem turns into one of accounting for a loss of energy when a neutron changes into a proton and an electron, and there is no exchange with another proton in the nucleus. Accordingly, in 1927 Wolfgang Pauli suggested that another particle was formed at the same time, and seven years later Enrico Fermi developed and extended the idea. Since no increase in mass nor any change in electric charge was called for, the particle could possess neither electric charge nor mass. The only thing to save it from being completely imaginary was its spin which, in order to conserve the angular momentum that appeared to be lost in the neutron decay process, was assigned a particular value. This peculiar particle was named the neutrino and, in spite of its almost ethereal quality, has been found of use not only in explaining every aspect of β-particle generation, but, since its apparent detection in 1956 during nuclear reactions observed in the U.S.A., also as a means of detecting reactions within the centres of stars, as will be mentioned in Chapter 8.

The neutrino and positron are not the only strange particles that have been devised to hold together the particle theory of matter, there are also transitory particles known as mesons. The need for these arose out of the problems within the nucleus to account for all the exchange phenomena between particles that form the binding forces of the nucleus itself. The theory of β-particle emission devised by Enrico Fermi using the neutrino as an integral part of the hypothesis, allowed all the energies involved to be computed, while the experimental evidence of collisions between protons and neutrons also provided a series of energy values. But trouble arose when it was found that the two sets of energy values did not coincide. To overcome this discrepancy it was suggested by the Japanese physicist H. Yukawa that where a neutron decays it splits into a proton and a negatively charged meson, with a mass some 200 times greater than the electron, and only after this does the electron form due to the decay of the meson which, in addition, produces a neutrino. The emission of a positron could also be explained by the decay of a proton into a positive meson and a neutron, and then the decay of the positive meson into a positron and a neutrino. Later, observation of nuclear particle reactions showed that Yukawa's mesons – the so called μ (mu) mesons – appeared to be more than a useful fiction. Yet the μ-meson did not, in fact, satisfy every observation and twelve years later, in 1947, Cecil Powell at Bristol and others adopted an interpretation of their observations suggested at this time by Bethe and Marshok: this interpretation suggested that there was an even heavier type of meson – the π (pi) meson – which could be negative, positive or without electric charge. A π-meson then decayed into a μ-meson and a neutrino unless it had a neutral charge, in which case its decay produced γ-radiation. Another stage was added to the disintegration of protons and neutrons.

The situation has changed since 1947, and unhappily the elementary particles have proliferated still further. A third kind of meson – the K-meson – has been created to account for the reactions between π-mesons and protons or between a proton and a neutron, and another neutral meson – the omega particle – is believed to exist: it decays into three π-mesons. As if the family of mesons were not enough, a new class of elementary particle – the hyperon – has been suggested. Such particles have a very transitory existence, of the order of a ten-billionth of a second (10^{-10} sec.), and are generated in the same way as K-mesons, by the collisions of π-mesons and protons or collisions of neutrons; they decay into π-mesons, protons, neutrons, other hyperons (which then decay as described) or, in one instance, into another hyperon and γ-radiation.

The reader may well be forgiven if he asks whether all these particles really exist and, if they do, whether they are the real elementary particles. The situation is complex but, as mentioned at the beginning of this chapter, it appears that if we are to hold a particle hypothesis for the nature of matter, and keep this hypothesis consistent with the fundamental laws that energy, matter, and momentum can never be lost but only distributed in a different way, then this plethora of particles must be accepted. They seem the one way to rationalize our observations. A fresh approach may completely replace the present complexities – the history of scientific thought would lead one to consider this likely – but at present there is no substitute. Some physicists are seeking a simplification, however, and trying to conceive of fundamental particles of a simple kind so that they may interpret the present explanations in some way other than a system of basic units of matter: whether they will be successful it is too early to say. Nevertheless, it seems significant that the hyperons and K-mesons have, at the moment, to be given what the nuclear physicist calls a 'strangeness' number to help describe their behaviour, even though the significance and meaning of this number is not yet understood. Some further elucidation and, presumably, simplification, would appear to be vital.

It was Paul Dirac who, in 1928, produced a theory of the electron that made it seem likely it should have an anti-particle counterpart: in other words that there should be a second particle just like the electron but with a positive instead of a negative electric charge. The discovery of the positron was, in fact, the discovery of just such an anti-particle, and it has led physicists to wonder whether there may be other anti-particles; whether, perhaps, every fundamental particle has its own anti-particle of the same mass but opposite electric charge. There is nothing in the Dirac theory to oppose this and in 1955, in a nuclear accelerator at Berkeley in California, nuclear particles were accelerated to a very high energy (some twenty-four million electron volts) and aimed at a piece of copper. Nuclear particles were dislodged and among them was what to all intents and purposes appeared to be a proton with a negative charge – an anti-proton – and this observation has been repeatedly confirmed.

This discovery is of more than academic interest; since now there seem to be anti-electrons (positrons) and anti-protons, one can conceive of atoms having a nucleus with anti-protons orbited by anti-electrons. The nucleus would have a negative charge and the anti-electrons a positive charge, otherwise the atoms of 'anti-matter' would be the same as those of ordinary matter. However, when an anti-electron meets an ordinary electron, it is known that the particles are annihilated and their mass transformed into an immense release of energy, evident as γ-radiation. In view of this, it has been suggested that there may well be anti-matter as well as normal matter, and that such anti-matter may be generated in the central regions of some or all galaxies. If this were so, then the anti-matter might be expected to meet normal matter, of which there is obviously a plentiful supply, and be annihilated. The energy released in such an annihilation would be enormous, but it has been suggested that this may be the only way in which the energy observed to be released in distant galaxies and in other objects far off in space, can be generated.

But if the energy released in the annihilation of matter and anti-matter is enormous, that released in nuclear fusion is also large. According to relativity theory* the quantity of energy released when one gramme (about one twenty-eighth of an ounce) of matter is annihilated, is some 20×10^{12} calories, or enough energy to raise the whole 300 thousand, million, million tons of water in the Atlantic Ocean 2°C. With nuclear particles, which have masses only some 10^{-24}, the energy generated is proportionally less but, on the other hand, there are vast numbers of these particles inside the stars taking part in the nuclear reactions there. In the Sun, for example, there are enough to generate sufficient energy in one second not only to boil the Atlantic Ocean but to evaporate the Earth. We must now enquire how this energy is generated and, to begin with, it will be best to commence with the proton-proton reaction, which has three stages.

The process commences with the collision of two protons (hydrogen atom nuclei) which are unstable, since both have a positive electric charge. A reaction therefore occurs and one proton turns into a neutron before the collision can be completed. The neutron combines with the other proton, forming a nucleus of deuterium, an isotope of hydrogen and sometimes called heavy hydrogen. The change of a proton into a neutron results in the emission of one neutrino and a positron. The process is an infrequent one; even in the Sun's central regions the chances that any two protons will meet one another is approximately one chance in fourteen billion years, but this is an average figure for two protons, and there are so many protons there in close proximity that enough encounters occur every second to permit energy to be generated in sufficient quantities to keep it shining.

The second stage of the proton-proton reaction happens when another proton combines, this time with the deuterium nucleus. This occurs quickly (average time six seconds), and a helium nucleus is formed, a γ-ray photon emitted, and 5·5 million ev of energy generated, compared with a quarter of this in the first stage of the reaction. The new helium nucleus can react in a number of ways, but the most likely third stage of the process is for two helium nuclei to combine to form a heavier isotope of helium (an α-particle), with the ejection of two protons and a γ-ray photon. The energy generated in this way is 12·9 million ev, but like the first stage of the reaction this is infrequent, two helium nuclei meeting and transmuting once every million years on the average. However, there are so many helium nuclei that this figure is apt to be misleading

* This is the famous equation $E = mc^2$, where $m =$ the mass of matter annihilated, $c =$ the velocity of light, and E the energy released.

and certainly the rate of generation of solar energy shows that many such reactions occur every second, even though the first stage has to happen twice to produce enough helium nuclei for stage three.

The carbon cycle is more complex. It has six stages, in the first of which the nucleus of a carbon isotope (carbon 12) combines with a hydrogen nucleus (proton) and transmutes into a nitrogen isotope with the emission of a γ-ray, after which the nitrogen isotope spontaneously disintegrates into a different carbon isotope (carbon 13). Recombination with more hydrogen produces a more stable isotope of nitrogen, which can then combine with hydrogen to form oxygen; but the oxygen isotope created is unstable, it breaks down into yet another nitrogen isotope which combines with hydrogen to make carbon (carbon 12 again) and helium, and bring the cycle to a close. Beginning with carbon and hydrogen, the process ends with the same isotope of carbon, but with an α-particle instead of the hydrogen nucleus. As in the proton-proton reaction, hydrogen is transmuted into helium with the emission of energy.

When stars leave the main sequence, their cores contract and become still hotter than before, and then there are other thermonuclear reactions that can occur utilizing the nuclei of heavier elements. Helium nuclei can combine to form beryllium, followed by a reaction between the beryllium and more helium to give carbon and emit radiative energy. Other processes in which oxygen, neon and magnesium are formed are also believed to take place until the internal temperature reaches some three billion degrees, when the nuclei of these heavier atoms will react among themselves, forming still heavier atoms up to iron. At this stage neutrons are generated in great profusion; neon and helium nuclei can produce great quantities of neutrons which, when created as in a supernova explosion, are captured by other nuclei to synthesize very heavy elements – indeed, this appears to be the way in which they are introduced into the universe if, as seems generally accepted, hydrogen was the basic element.

In radioactivity electrons are emitted, and this β-decay process arose, as was mentioned, from the breakdown of a neutron into a proton. It has been suggested that, in the centres of stars close to the end of their life, where the densities of nuclear particles and electrons is very great indeed, an inverse β-decay process occurs. Here protons and electrons combine to form neutrons and neutrinos. This process cannot continue without limit since too many neutrons in a nucleus lead to instability and nuclear fission would take over, with neutrons and protons being ejected until stable nuclei of lighter elements are formed. With this in mind, astrophysicists argue that in massive stars, originally heavier than the Sun – theory gives the value 1·44 times heavier – a complex process occurs, leading to the formation of a neutron star.

There is some argument about the details, but in general the picture appears to be that by the time heavy nuclei are being synthesized in the core, there has been considerable contraction, this contraction increasing each time a new type of thermonuclear synthesis occurs: when at last the inverse β-decay procedure begins it continues for a time absorbing radiation which it obtains from the immense contraction. The temperature rises, electrons and positrons are produced in great abundance, but they are in a restricted space and there will be many collisions with the result that they are annihilated. Pairs of neutrinos and anti-neutrinos will also be formed as a consequence of this annihilation, and energy is dissipated at an immense rate by the ejection of neutrinos. The star core now collapses, becomes hotter, more neutrinos are generated and emitted, and the collapse proceeds still further. Some fission of iron and other nuclei will now probably occur, producing the lighter element helium and the ejection of still more neutrons. The fission process here involved requires energy and since the star cannot be expected to freeze, it must contract still more, dense though it already is. However, since this contraction is to provide energy for the fission process, the energy thereby generated is not available for raising the internal temperature of the core.

This stage of the contraction is thought to be a catastrophic process – a virtual implosion of the star that takes something of the order of one second and no more. If a stellar core collapses like this, the surrounding material will collapse also, and this will increase in temperature as it does so. Because of the heating which also occurs in something like one second, the oxygen shell near the core (formed originally from the synthesis of helium into heavier elements) will burn up. This is observed as a supernova outburst. As far as the core itself is concerned, once it has reached a density of about 100 tons per cubic inch and the temperature is some six billion degrees, the inverse β-decay process accelerates and the core continues to collapse. How long the collapse continues is uncertain, but it would seem likely to cease only when the density of the neutron material reaches the astounding figure of 100 million tons per cubic inch. The great heat (6×10^9 °K) would cause a prodigious ejection of neutrinos, and energy would be lost at so high a rate that the neutron core would remain in its super-condensed state, since there would be no energy available to make it rebound a little after its collapse.

The outcome of all this is what has become called a neutron star. At the centre would be a core about ten miles in diameter, composed

primarily of neutrons in abnormally close proximity and whose behaviour is restricted – in a degenerate state, in fact – with perhaps a few
closely packed electrons, protons and hyperons as well. Outside the core
a thin shell of atomic nuclei would lie with degenerate electrons closely
packed amongst them and moving with immense speeds: this shell is
thought to be about half a mile thick. Beyond the shell should be
another, no thicker than some forty feet, similar to the half mile thick
shell in composition, although the electrons here would be moving more
slowly. A final shell of iron and lighter elements, but with neither hydrogen nor helium, would form the outside of the star. Its temperature
should be in the region of ten million degrees but its thickness a mere ten
or twelve feet. The total diameter of such a star works out at some eleven
miles and would be ordinarily undetectable by visual observers unless
it lay within the solar system. Theoretically, it could be observed by its
X-radiation which would be emitted because of the extraordinarily high
temperature of the tiny object, or due to radiation of other kinds (p. 110).
It is also considered possible that a white dwarf could decay into a
neutron star, but in this case the change would be less dramatic.

Astrophysical theory has considered the formation of objects even
denser than neutron stars. According to the Russian astronomer Viktor
Ambartsumian hyperon stars may exist, for although under normal conditions hyperons have a very transitory existence, where the density of
neutrons and protons is so great and their energy so small (degenerate
conditions), hyperons are prevented from decaying into neutrons or
protons. Mesons would also be formed but, because of the density and
energy involved, they too would remain stable. Such hyperon stars
would, Ambartsumian conjectures, possess a hyperon core in which
most of the mass is concentrated, a thin shell of degenerate neutrons
with a few electrons and protons, and an outer region with normal
nuclei and electrons. Beyond this outer region there seems likely to be a
very thin shell of normal atoms. The density of a hyperon star can be
judged from the fact that although it is probably as massive as the Sun,
its diameter is no more than eight miles, or 100,000 times smaller.

According to the general theory of relativity, light (and other electromagnetic radiation) should be deflected when passing a massive body,
and at the total solar eclipse in 1919, observations were made of stars
close in the sky to the Sun, rendered visible during the few minutes when
the light from the photosphere was cut off. Their images were found to
be shifted as the theory had suggested. Neutron and hyperon stars display great density in a small space, and their strong gravitational fields
ought to curve starlight passing close to them by an extraordinary

amount; indeed it has been suggested that they would act like 'gravitational' lenses. If this does happen, then the light of distant stars would be formed into images on the surface layers of a neutron and hyperon star and these 'ghost' images could, in theory, be observed visually. No clear and unequivocal observations of this kind have so far been made, although there would seem to be other evidence – X-ray, radio and visual radiation – for the existence of such stars (Chapter 6).

One of the techniques now much used by the nuclear physicist for investigating the behaviour and discovering the existence of fundamental particles, is to accelerate protons and electrons to very high velocities – velocities that approach the velocity of light. To obtain such high speeds the particles are raised to a considerable velocity by passing them down a long evacuated tube fed with a very high voltage, and then injecting them into an evacuated circular chamber. The velocity is now raised still further by the application of high frequency pulses of very high voltage synchronized with the motion of the particles, and in such a synchrotron, the energies of the particles may reach Giga (10^9 ev) electron volts. When electrons are caused to move in circular paths inside such a machine, through which there is a strong magnetic field (to keep the motion of the particles in a curved path), they emit radiation. The radiation passes away at right-angles to the path (figure 12). As this

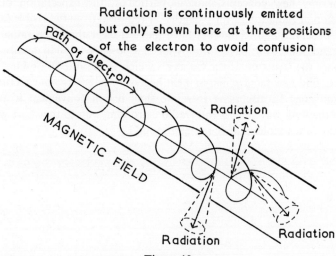

Figure 12

Synchrotron radiation. The parallel straight lines represent a magnetic field. The radiation is continuously emitted at right angles (i.e. tangentially) to the path of the spiralling electron.

is a natural phenomenon due to the immense speeds involved and the curved path of the electrons, one might expect this to occur in space where there are strong magnetic fields and where electrons are accelerated to very high velocities. The wavelength of such 'synchrotron' radiation depends both upon the strength of the magnetic field and the velocity of the electron; it may be at radio wavelengths or, if field and velocity are very great, X-radiation.

X-rays may also be generated in other ways. If, for example, protons or electrons become accelerated with immense velocities, it appears possible that they could penetrate the outer electron shells of heavier atoms like iron and eject one or more electrons from an inner electron shell. Such inner shell ionization would release considerable energy that would be manifest as short-wave X-radiation. Again, if a photon of starlight is struck by a very high velocity electron, the electron's path will be altered and X-rays will replace the longer wavelength starlight. Lastly, in a hot gas such as exists in various parts of the Galaxy (and in other spiral galaxies) some free protons may be expected to exist, and the passage close to them of very high speed electrons will also produce X-radiation.

The study of nuclear energy and of the generation, annihilation and behaviour of the so-called fundamental atomic particles, has brought to astronomy new concepts and, in particular, the expectation of the emission of radiation in wavelengths far removed from the visible range. New observational techniques have made it clear that such radiations exist and, moreover, have not only made it possible to fill in details – for instance the structure of the Galaxy – which were hitherto obscure, but also to find new facts which make it necessary to enlarge our picture of the universe. In discussing these we shall find it convenient to move from the long wavelength end (radio astronomy) up to the short wavelengths (ultra-violet and X-ray astronomy).

CHAPTER 6

Radio Astronomy

It was in 1932 that Karl Jansky, investigating the crackling sounds to be heard in short wave radio transmissions at fifteen metres, found that a steady hissing noise was also to be heard. This hissing noise reached a peak when his aerials were pointing to a region of the sky in the constellation Sagittarius, and he reported his result. Scientifically it aroused little interest except in an American amateur radio enthusiast, Grote Reber, who built himself special equipment and finally, using a wavelength of 60 cm, was able to detect radio emission from the Sun, and the constellations Cassiopeia and Cygnus, as well as from the Sagittarius region. Reber announced his results in 1940 and 1942, while short wave radio techniques were being considerably improved in the wartime development of radar, and although little notice seemed to have been taken of them at the time, solar radio radiation was found to affect radar efficiency. Some measurements and preliminary investigations with radar equipment were made by G. C. Southworth in the United States and James Hey in England, but it was only after the close of hostilities in Europe that the matter was taken up in earnest, first by radio engineers and later by astronomers.

The first requirement for a radio investigation of the universe is a directional aerial or antenna. For ordinary (short, medium and long wave) broadcast reception an antenna does not require to be directional, and a piece of wire is entirely satisfactory. However, this is not so in much shorter wave bands and in ultra-high-frequency television transmissions: indeed from wavelengths of some twenty metres downwards a directional antenna is vital. It can take many forms, and in Jansky's case was a square wooden framework around which wire formed a square loop, but the reception of wavelengths half as long is better served by the dipole – a pair of metal poles joined by insulating material at the centre, giving a total length equal to half the wavelength to be received. The well-known television receiving antenna is of this kind. Where there is a need to increase the strength of the incoming radio signals – in other words, to cause the antenna to collect as much radiation as is practicable – reflectors may be used. The simplest form is a parallel bar, a little longer than the dipole, while to increase directivity, a number of parallel bars, a little shorter than the dipole, may be placed in front of it. Alternatively, a reflector may be flat with a surface of wire netting or some similar mesh, provided the holes are very much shorter than the wavelength to be received. Antennae of this kind are in use at the Mullard Radio Astronomy Observatory at Cambridge, England; as also are aerials with a 'corner reflector' – a dipole, or series of dipoles with a right-angled framework fitted behind, rather as if one had cut a flat reflector in half and joined the pieces perpendicular to one another.

In radio astronomy the signals to be received are very weak by terrestrial broadcast standards and, in addition, since the radio astronomer clearly wishes to receive signals from the most distant objects he can, he has to use directional antennae of as great a sensitivity as possible. The antennae of the larger radio telescopes are therefore all built with parabolically shaped reflectors, whose action is analogous to the main mirror of an optical reflecting telescope – they collect the radio radiation and reflect it to the dipole at the focus of the parabola. The parabola itself may consist of no more than a framework covered with narrow mesh wire or, if the wavelengths to be observed are less than some ten centimetres, with a sheet of metal.

Just as the shape of an antenna can take many forms, so can the way in which it is mounted. Since the antenna is directional it will receive the strongest signals from sources directly in front of it, and it might be expected to require an elaborate mounting like an optical telescope. This can be an advantage for some kinds of research, and the most highly developed form is the equatorial mounting, where the antenna is pivoted on an axis that lies parallel to the axis of the Earth: rotation in one direction only then permits it to follow the radio source as it appears to move across the sky owing to the Earth's diurnal rotation. Ideal though this mounting is – and it is almost universally adopted for optical telescopes – there are cost and engineering limitations for radio telescope antennae of very large size. As will be clear in a moment, antennae have to be far larger than any optical mirrors, with diameters measured in feet rather than inches, and for this reason many are mounted in the 'altazimuth' form, where the axis of rotation is vertical, not inclined. The world's largest steerable antenna, the 250-foot diameter Mark I radio telescope at Jodrell Bank in Cheshire, has an altazimuth mount and this is the form most widely adopted elsewhere for an instrument that can be directed to a series of points in the sky during an observing session. However, the radio telescope can be used when there is cloud and mist coverage – conditions that would prevent any optical observing whatsoever – and it is therefore feasible to construct very large antennae that can be moved in a north-south direction only, and utilize the Earth's rotation for an east-west movement. In the 1000-foot-diameter antenna at Arecibo on the northern coast of Puerto Rico, the reflector is fixed in a natural bowl-shaped valley and the Earth's rotation provides the east-west movement, but since it is impracticable to tilt the reflector, movement in a north-south direction is achieved by moving the dipole which is mounted in a special movable cradle: even so, observation can only be made within 20° of the zenith.

The resolution of detail with a radio telescope depends on the size of its reflecting system and is proportional to its diameter, as is the case with an optical telescope. However, as the radio wavelengths are greater by a factor of at least 10^4, to obtain equivalent resolution (to fractions of a second of arc), parabolic reflectors would need to be 100 miles or more across, and this is obviously impossible. For much observing, radio astronomers have to be satisfied with a resolution measured in minutes of arc. The 250-foot Jodrell Bank instrument has a resolution of some 15' at a wavelength of 30 cm, while the 140-foot radio telescope at Green Bank in West Virginia that is designed to operate at a wavelength of 3 cm has a resolution of 2'. Much useful work has been done with what is poor resolution by optical standards, but for mapping radio sources some improvement is required, and the radio interferometer has become widely adopted. This is based on the same principles as that of the stellar interferometer mentioned in Chapter 3, and at its simplest* uses two antennae separated by as large a distance as is convenient, the signals being fed into an electronic mixing unit and then passed to the radio telescope receiver. The distance apart of the antennae gives a resolution equivalent to a fixed reflector of diameter equal to this distance.

The first radio interferometer of this kind was constructed by Sir Martin Ryle and D. D. Vonberg at the Mullard Observatory in 1946 for the purpose of examining spots of solar radio emission – it was a pair of simple dipole arrays with wire netting reflectors – but since then there have been many variations. For instance, Bernard Mills, an Australian radio astronomer, has devised an interferometer utilizing two long lines of cylindrical or parabolic reflectors set at right-angles

---

* To be precise, the simplest of all interferometers was constructed by J. L. Pawsey in Australia, who mounted an antenna on a cliff edge and, with great ingenuity, utilized the direct signal received in the antenna and that which was reflected from the surface of the water.

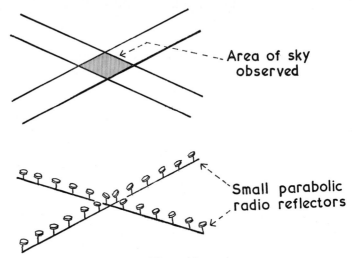

Figure 13
The principle of the Mills Cross type of radio interferometer which utilizes
two long rows of antennae at right angles. These can be either long cylindrical
antennae or a series of parabolas (as shown in lower picture). The area of the
antennae and that of the actual area of sky covered when the signals received
by the antennae are mixed, is illustrated in the upper part of the figure.

across each other (figure 13). With this, the Mills Cross interferometer,
surveys can be made over a small square section of sky, resolution
depending again on size – in the case of this interferometer on the
lengths of the arms – and on wavelength: working at a wavelength of
$3\frac{1}{2}$ metres and with crossed arms 1500 feet in length, its resolution is 50'.
Other crossed interferometers have been constructed, both in Australia
and elsewhere, some using two series of parabolic reflectors (in place of
wire reflectors) for shorter radio wavelength observations. Large parabolas at different observatories have also been used interferometrically,
Henry Palmer using a base-line extending 80 miles from Jodrell Bank,
while intercontinental links are being arranged.

The Mullard Radio Astronomy Observatory has used interferometers
with four aerials spaced in rectangular formation, and from these Martin
Ryle has developed a technique of observing known as 'aperture synthesis', a technique that has no equivalent in the optical field. Instead
of building two long antenna arrays at right angles, he constructed one
in an east-west direction; in the north-south direction, only a short
stretch (200 feet) of array was built, though it was mounted on a
specially constructed railway track 1000 feet long. The length of the

east-west fixed aerial was 1450 feet, and the equivalent of a telescope
with a rectangular area 1450×1000 feet was synthesized by making
observations with the north-south aerial in successive positions. A complete set of observations takes time and the complex analysis must be
handled by a computer, yet with this instrument a resolution of 1′ at
metre wavelengths has been obtained. More elaborate still is a three
element aperture synthesis telescope that Ryle has constructed, with
three equatorially mounted 60-foot-diameter parabolas, whose distance
apart may be anything up to one mile in an east-west direction. The rail
track is correct to within one eighth of an inch over its entire length, and
observations as if with a one mile diameter parabola can be made,
utilizing the rotation of the Earth to sweep the east-west track round
one revolution per twenty-four hours. Observing is carried out at two
wavelengths simultaneously – these are at 70 cm and 20·7 cm, and at
the latter wavelength the instrument possesses a resolution of 23″. For
very short radio wavelength observations, of the order of a few centimetres, the ordinary parabolic reflector or interferometer is satisfactory,
but for longer wavelengths interferometers and aperture synthesis are
vital. Admittedly the total power of the radio signals received in an
interferometer or by aperture synthesis can never be as great as would
be received from a complete parabola that fills, say, a diameter of a
mile, but at the longer wavelengths, particularly in the metre range,
signal strength is greater than at the shorter wavelengths, so the disadvantage is not as great as it might at first sight appear. Moreover a
very large parabola is impracticable and the resolution that can be
achieved by, for instance, aperture synthesis is well worth the comparatively slight loss in signal strength.

The antenna is the more spectacular part of a radio telescope, but it
is no more than a part: the signals it receives must be detected and
amplified by a radio receiver before any use can be made of them. Since
these signals are always very low in intensity compared with those used
in radio communication, special kinds of radio receiver have been
devised. The problem really arises because of the 'noise' generated within
a receiver by the components, of which the electrical resistances are the
worst offenders. They generate a hissing noise and, unhappily, this is
similar to the hissing noises received from a radio source in space,* so
with ordinary receiver techniques the number of extra-terrestrial radio

* That the radio astronomer does not turn his signals into sound is irrelevant. He
records signals on a chart with an automatic pen recorder (in which the pen is moved
by an amount proportional to the hissing signal received), on to punched tape for a
computer, or by both methods. The hissing noise is what the signals would give if
played over a loudspeaker and is a useful way of describing them.

sources that could be observed would be very few, since most would be swamped by receiver noise. There is also the problem of a background noise from the sky, as observations are made through the atmosphere which is not completely transparent electrically, but at frequencies above 200 MHz (wavelengths shorter than 1·5 metres) it is less of a trouble and it is receiver noise that is the main difficulty.

Two major advances in radio receiver design have been devised to overcome this problem as far as possible: the maser and the parametric amplifier. The maser, so called because it is a device for *micro*wave *a*mplification by *s*timulated *e*mission of *r*adiation, is in essence a receiver in which radio energy is stored and released when an incoming radio wave of the same frequency stimulates it to do so. The storage element is a crystal into which radio energy is 'pumped' by a radio oscillator working on a higher frequency, and it enables an amplifier to provide high amplification at a specific frequency chosen for the observations in question.* Its advantage is obvious, since a radio signal from a source passing across the antenna will stimulate the crystal into emission. There are practical difficulties in using a maser, as the crystal (usually of ruby) has to be kept at a very low temperature using a coolant such as liquid helium, and requires an especially stable magnetic field to be maintained around it, while the radio power fed to it by the oscillator must be just the correct amount. Therefore more use is being made of the parametric amplifier.

The name refers to the fact that one of the radio component values or 'parameters' of the receiver's amplifying circuits is regularly changed by an oscillator and stores energy, which it can then use in amplifying an incoming radio signal. The component used is a capacitor (essentially a device of two metal plates that can store an electric charge) and the advantage of using energy from it to amplify a radio signal fed into the circuit is that it produces little noise of its own, and can thereby reduce the background noise to a degree impossible in an ordinary amplifier.

Two basically different observing techniques have been used in radio astronomy – that of radar, whereby radio pulses are transmitted and reflected back from distant objects which need not themselves emit any radio radiation at all, and the more straightforward method of receiving radio radiation emitted from celestial objects in space. The latter is the main concern of radio astronomers, but radar techniques have their own

---

* The maser principle may also be used for obtaining very intense beams of light energy – the principle is similar and the device is then known as light amplification by stimulated emission of radiation, or a laser – an instrument far better known than the maser, on which it is based.

special use for astronomically nearby objects, and it will be convenient to consider them first, particularly as they were used for some of the earliest radio astronomical work by Sir Bernard Lovell and James Hey.

Radar techniques were applied to the study of meteors, which are not easy to observe optically with a high degree of precision. They are sporadic and only on special occasions – when the Earth intersects a meteor stream – are there showers when there is a statistically satisfactory number of meteors present to determine their paths with desirable accuracy. All the same, the optical observer has achieved considerable. success, but there was still one outstanding problem that faced him after the end of the Second World War, and this was the question whether most meteors were part of the solar system or intruders from interstellar space. Some meteor showers were clearly connected with the orbits of periodically recurring comets, being, in fact, debris of cometary material spread along a comet's orbit, but the question had remained unanswered as far as sporadic meteors were concerned. The crucial question was to determine not only the orbital direction of meteors but also their velocity, since if the velocities were higher than a particular value this would indicate that they were moving too fast to be constrained in an elliptical orbit and might have an hyperbolic path: the latter would show that they were temporary visitors to the solar system.

Photography appeared one way to solve the problem, but was radar another way? When meteoric material enters the terrestrial atmosphere, its velocity is of the order of at least four miles per second and this is sufficient to cause the lump of rock or metal that composes the meteor to heat by friction with the surrounding air. Indeed in nearly every case, the material is evaporated before it reaches the ground, but as it burns, it leaves behind a trail of glowing ionized air particles that lasts for a fraction of a second or so. It is this trail that is visible to the observer (except in the case of a large meteor or 'fireball' when hot glowing gases may also be seen surrounding the head), and the ionized gases act as an excellent reflector of short radio waves. In 1945, when James Hey and his colleagues at Malvern began to investigate the matter, photographic techniques were not very satisfactory but many successful radar measurements were made. Later, especially in the hands of Frank Whipple at Harvard, photographic methods were improved, and these have produced similar results to those of Hey in showing that there appear to be no hyperbolic velocities. However, the radar technique, which could be continued in cloud and, more important, in daylight when no visual or photographic observations could be made, was pursued particularly by Sir Bernard Lovell at Jodrell Bank and it became clear that the Earth received many daylight streams and many daylight sporadic meteors as well. Lovell's investigations confirmed that the velocities of all these daylight arrivals were less than hyperbolic and it is now accepted that meteoric material is an inherent part of the solar system.

The distances over which radar can be used are clearly limited by the fact that radio pulses have to be emitted, reflected and returned and, although these travel with the velocity of light, it is impracticable to send them even as far as the nearest star. Moreover, to send pulses out and back over such a distance (four and a third light years each way) is not technically feasible at present due to loss of energy. The reason for this is that even a radio pulse will spread out as it travels into space and back, and some of its energy will be absorbed on reaching its target: in addition, as the target will be a spherically shaped object, such radiation as is reflected will not all return to the observer. As a consequence radar observations are confined to the solar sytem, and in general fall into two classes: those for determining position, and those that can provide some indication of the nature of the lunar or planetary surfaces. The position determinations are useful because there is considerable difficulty in making distance measurements by optical means, and radar determinations can provide values with a smaller margin of error. Radar echoes have been obtained from the Moon, from Mercury, Venus and Mars, and from the Sun, the reflection of the pulse in the Sun's case occurring in the corona. These radar echoes represent a considerable achievement when it is realized that the sensitivity required to receive echoes from Venus when closer to the Earth, for instance, is some ten million times that required for similar echoes trom the Moon. All the same echoes from the Sun, the Moon and from Mercury, Venus and Mars have not only been successfully received, but have also led to a better determination of the distance of these objects.

Radar echoes from the Moon have also been used to investigate its surface with some success. The radio observatories at Jodrell Bank, Sydney, Washington, Arecibo, Goldstone, California and Tyngsboro, Massachusetts have all taken part in the investigations, which have shown wide variations in the strength of the echoes received. Some of this variation is due partly to the wavelength used, and partly to the fact that the Moon's rotation on its axis is not exactly in step with its orbital motion round the Earth, the latter causing the Moon to appear to swing its rotation a little west or east as far as a terrestrial observer is concerned. This 'libration' of the Moon's rotation may, at maximum, cause the surface to appear to move some two and a half miles per hour across the line of sight and such a relative motion causes multipath reflections, some of which destructively interfere with one another and so cause fading. Nevertheless, at Jodrell Bank, this phenomenon of fading was utilized to make some assessment of the surface which gave evidence of appearing 'smoother' than visual observations had led astronomers to suspect. These observations have been confirmed by the American Surveyor space probes that have soft-landed on the Moon, and the Russian probes, Luna 9 and Luna 13 particularly. American radar observations, especially of the surface of Venus, which cannot be seen visually because of the thick cloud layer that covers the planet, have also provided an indication of its nature. They have shown that there appears to be some high ground in certain areas, and although this may not seem a particularly epoch-making result, it is significant since there has been argument about whether the surface is perhaps entirely covered with water. The soft-landings in 1967 and 1969 of the Russian Venus probes have confirmed the existence of a solid surface.

A new technique of displaying radar results, developed at the Lincoln Laboratory of the Massachusetts Institute of Technology at Tyngsboro, has also resulted in radar being used to provide a picture of a planetary surface, and in the case of the Moon, the technique at present gives results as good as those obtained with an optical telescope. Professor Shapiro at Tyngsboro confidently expects this method to surpass optical results in the forseeable future.

Shapiro and his colleagues have also developed a method of using the Doppler shift given to radar echoes by reflections from the opposite limbs of a planet that is in rotation, to provide a value for the actual period of revolution. The limb approaching the observer causes a rise in frequency of the echo, and the limb that is receding gives a drop in frequency; as a result a rotation period of about fifty-nine days was found for Mercury in 1965 by Gordon Pettengill and Rolf Dyce at Arecibo – a very different value from the period of eighty-eight days suggested by visual observers. A period of 243 days has also been determined for Venus, and its rotation has been found to be retrograde (from east to west instead of the usual west to east direction of the other planets).

In addition, Shapiro has used radar observations as a test for Einstein's general theory of relativity. Visual observations made in 1919 and later showed that a shift of starlight envisaged by the theory could be detected when stars appeared close to the Sun's limb during a total solar eclipse. But these results have always required some independent confirmation. Shapiro pointed out that not only should there be this deflection of starlight due to the Sun's gravitational pull on the photons, but also that the theory would lead one to expect a retardation of their velocity. Using echoes from Mercury when on the further side of its orbit, so that the outward and return radar pulses passed close to the Sun, and comparing these with similar echoes when Mercury was close to the Earth, such a delay was in fact observed. The precise kind of measurement involved in this delicate and elegant experiment may be judged from the fact that the observed delay amounts to no more than 200 millionths of a second, a fraction of one part in ten million of the total delay between the outward and return signals.

Radio astronomy proper, the receipt and analysis of radio radiations emanating direct from space, is not confined in any way by distance, any more than visual astronomy, and has been applied from the study of the Moon to the most distant objects. The lunar investigations have arisen at wavelengths of around some 8 mm and have been concerned with heat radiation from the surface: they are, in fact, an extension of the infra-red radiation studies mentioned in Chapter 3. The Moon's heat radiation is caused by the heating it receives from the influx of solar radiation, and it might be expected to show considerable changes at times when sunlight is irradiating the surface compared with those when it is not, since the Moon possesses no atmosphere to retain any heat, as has the Earth. Yet, surprisingly, the total range of temperatures between the extremes of the maximum period (two weeks) of solar heating and of maximum cooling (two weeks) was no more than 60°C, and when, during the period of the observations, the Moon was eclipsed by moving into the Earth's shadow, again no significant change was noticed. These results were different from those of longer wavelengths (long infra-red) and clearly must be derived from radiation below the surface 'ground' of the Moon, although it seems that the insulating layer penetrated in this way is unlikely to be more than about two inches. The observations have also made it clear that the surface is not solid rock, as the Surveyor space probes have also proved.

As far as the planets are concerned, all emit thermal radiation, some of which falls in the radio spectrum. This radiation is weak, of very short radio wavelengths – some 3 cm – and very sensitive receivers are necessary. The results obtained are difficult to differentiate from other unwanted radiation at this wavelength but, nevertheless, they give values for Mars and Jupiter similar to those obtained by infra-red measuring equipment. Venus, however, presents a different picture, as the radio temperature is more than twice that given by the infra-red. This difference has now been shown (by the Venus 4 probe) to be due to the radio emissions coming from below the thick cloudy outer layer, and shows that what the radio telescope observes is often quite different from what is seen by an optical telescope. Jupiter's temperature is what would be expected from a planet at a distance of some 483 million miles from the Sun and, presumably, even the short radio waves do not arise from far below its immensely thick atmospheric covering, but Jupiter does present the unexpected phenomenon of sporadic outbursts of radio radiation at a wavelength of around fifteen metres and between eleven and seven metres. Since the planet does not radiate on its own account, any more than any other planet, and the radiations appear to come from one particular part of its surface, their cause is difficult to determine. To begin with radio astronomers considered whether these could be due to gigantic lightning flashes during thunderstorms on the planet, but a detailed analysis ruled this out, and some other explanation has had to be sought. It has therefore been suggested either that the effect is spurious and due to some quite different but unknown cause, or that the bursts of electrical activity are due to particles ejected from the Sun being trapped in the magnetic field of the planet. At present the question is still undecided, but if it does prove to be due to the behaviour of charged atomic particles, it is a form of synchrotron radiation (i.e. due to circular paths of the particles in a magnetic field) and, since it is not generated as are many radio waves in space by thermal conditions, it must be classified as non-thermal radiation – a term which is useful to the radio astronomer as will now become evident. More recent evidence obtained at the Owens Valley Radio Observatory in California by G. L. Berge and R. B. Read of radio emissions from Saturn may tend to strengthen the view, since these signals appear also to be atmospheric in origin.

The Sun, as already mentioned, is a source of radio radiation. Some of this is thermal in origin, as might be expected, but some is certainly non-thermal, and both types have been carefully studied since the Sun is close enough for all kinds of radiation to be observed, whereas other stars are too far away for this to be so. To consider the thermal radiation first, this has two components, one which periodically fluctuates in synchronization with the eleven year sunspot cycle, and one that varies over no more than weeks and is connected with the actual appearance of sunspots. The radiation connected with the eleven year sunspot cycle is similar to, and the radio extension of, the continuous spectrum that comes from the photosphere, but the area of the Sun from which the radiation comes varies with wavelength. At centimetre wavelengths the area of the 'radio Sun' is similar to that of the visible Sun, but at longer wavelengths in the metre range this is not so, and the radio Sun in such wavelengths appears larger than the visual image, at five metres being twice the diameter. This has been interpreted as meaning that the different wavelengths are emitted from different levels above the photosphere, and the interpretation is confirmed by the fact that at metre wavelengths the radio Sun appears elongated, spreading out sideways along the equator, the elongation being at its greatest at sunspot minimum and almost unnoticeable at sunspot maximum. This peculiar effect is equivalent to that observed visually at total eclipses when the corona is readily seen, for as previously mentioned, at times of sunspot minimum, the corona is elongated in just the same way, but almost evenly spread round the disk at times of sunspot maximum. The eleven year period or 'quiet Sun' radio radiation component has made it clear that the solar corona extends out much further from the Sun than can be gauged from any visual observations, and it now seems that it exists in a very tenuous state for many millions of miles beyond the photosphere.

The sunspot-linked thermal component is most noticeable on centimetre and decimetre wavelengths. Ordinarily the solar corona is transparent to these wavelengths but over the region of the photosphere where sunspots are active, the corona becomes denser than usual and is heated to a greater degree, which causes it to emit the thermal radiation observed. There is also an emission closely connected with sunspots but shorter in wavelength than ten centimetres, and this may also be a thermal emission within the chromosphere rather than the corona. There is some doubt about this since there is a possible non-thermal explanation for it, the so-called gyromagnetic emission of super-thermal (very fast moving) electrons, an effect which is due to the rotation of high speed electrons in a magnetic field and analogous to synchrotron radiation described in the last chapter.

In addition to the two thermally generated kinds of radio radiation. the Sun also emits a considerable amount of non-thermal radiation, This may be generated as synchrotron radiation, as gyromagnetic emission, by the Čerenkov process, or by plasma oscillation. The first two methods have been mentioned but something must be said of the other two. Čerenkov radiation (named after its discoverer) is that caused when an atomic particle enters a substance with a velocity greater than the velocity that light possesses in that substance. This does not mean that the particles themselves are travelling at velocities greater than light – the theory of relativity rules out that possibility – but only greater than the light inside the substance, and as the velocity of light is reduced in any substance from the value it possesses in a vacuum, this is nowhere near so unusual as it sounds. Čerenkov discovered that when such a very high energy particle enters the substance, it is immediately decelerated, and its loss of energy is evident as radiation. Over the Sun much of the gas is ionized, and such an electrically charged gas or 'plasma', as the physicist calls it, would appear a highly likely substance for the generation of Čerenkov radiation at radio frequencies.

The term plasma may indeed seem strange to a reader more familiar with the medical use of the word, but is used because an electrified gas behaves in a magnetic field in a way similar to a plastic (in the sense of mouldable) material. Such a plasma contains free electrons, detached from their atoms, and these electrons are in continual motion among the protons in the gas. When one moves close to a proton it will momentarily move in a curve round the proton and in doing so emit radiation. This is one source of radio radiation from a plasma, but it is of thermal origin since this electron motion is due to the temperature of the plasma. There is a non-thermal component, however. The non-thermal generation of radiation from a plasma is thought to be due to a disturbance within the ionized gas that causes a series of oscillations of the electrified particles: these oscillations cause, for instance, the electrons to move away from the protons, then return, overshoot their original position, move away again, and so on until the oscillation dies down. When the electrons are displaced, one side of any layer of the gas will possess a negative charge (owing to the surfeit of electrons there) and the other a positive charge (due to the excess of protons), this will then change as the oscillation proceeds, and oscillating charges negative-to-positive and positive-to-negative will be set up. These oscillations will give rise to radio radiation.

Of the non-thermal radio emission from the Sun, there are a number of different kinds, but all are associated with sunspot activity and with the development of flares of bright hot gas that erupt in areas close to sunspots or even at the edges of spots themselves. All the non-thermal emissions are transitory, ranging from a few seconds duration to a few hours, and although there are still many details yet to be discovered, the general situation seems to be that at the explosive start of a flare, when visual hydrogen emission occurs, radio emission commences at high frequencies (short wavelengths) and then rapidly drifts to lower frequency (longer wavelength), radiation lasting a few seconds at each wavelength. These radio bursts appear to be caused by plasma oscillations probably produced by Čerenkov radiation with its accompanying particle deceleration, the oscillations starting close to the photosphere and moving upwards into the chromosphere and then into the corona. A similar drift in wavelength occurs after the first burst, but the drift now takes longer

– minutes as compared with seconds – and is once more due to plasma oscillations but, this time, caused by a shock wave rather than a deceleration of particles emitted from the Sun. A cloud of ionized gas actually escapes from the Sun at this time, later reaching the Earth and causing sudden deflections of magnetic equipment – the so-called 'geomagnetic storm'. Continuous radio emission lasting minutes also occurs on metre wavelengths, due possibly to electrons (synchrotron radiation), although variations within the period make it a little difficult to see how such an explanation can really account for the phenomenon. Sometimes a long, strong burst of radiation on wavelengths from millimetres up to metres occurs, and is accompanied by another that probably arises in the corona and moves out to a distance of some two million miles from the Sun. After these two bursts are over, radiation at metre wavelengths continues irregularly and originates some distance from where the flare occurred: its cause is at present unknown.

Such radio frequency emissions are important to the astronomer since most have no analogous optical effects – visually only a flare is seen and nothing else – and provide an important means of learning what is happening above the photosphere. In particular they are a source of evidence of the emission of atomic particles from the Sun, and if, as is likely, this is a typical property of other stars, it is bound to be significant in questions of the behaviour of other objects in space, where high energy atomic particles play a considerable part in causing all kinds of phenomena. Certainly it is clear from visual observations and other radio observations, particularly those made in Australia and at Jodrell Bank, that some M dwarf type stars show very rapid fluctuations of brightness that may only last a few minutes, and seem to be due to giant flare eruptions. Many of the stars involved are members of binary systems and it has been suggested that in these systems the two stars may be very close together, in which case it may be that the strong gravitational pull between them which will distort the stars so that they are oval, can also draw off material in a flare disturbance. The fluctuations seem certainly not to be due to any eclipsing effect of one member of the binary system on another – the period of fluctuation is of the wrong kind for this explanation.

In general the radio emissions from stars are unobservable because of their distances and the weak radio signals they can be expected to emit, and although the term 'radio star' is still sometimes used, it refers to a source of radiation that is far from what the optical astronomer would call a star. But if radio astronomers cannot observe many true radio stars, there is certainly plenty of material within the Galaxy that emits at radio wavelengths, and of this clouds of hydrogen form a very important part. Hydrogen, when in a normal, non-ionized state, emits as we have seen at ultra-violet, visible and infra-red wavelengths. It appeared to the Dutch radio astronomer Hendrik van de Hulst and the Russian I. Shklovsky that hydrogen might also radiate in the radio wavelength region if one considered the possibility that the electrons could spin either in the same sense as the nucleus or in an opposite direction. This could occur in the cold regions of interstellar space where hydrogen atoms might, for the most part, be expected to be separate, and since there is a difference in energy between hydrogen where the spins are in the same sense and where the spins are otherwise, a change in spin should give rise to an emission of a photon of energy. Theory showed that the emission should occur at a wavelength of $21 \cdot 1$ cm. Van de Hulst and Shklovsky came to their theoretical conclusions in 1945 and 1947 respectively, but it was not until 1951 that American, Australian and Dutch radio astronomers succeded in observing this 21 cm emission. This was a very important step forward because, in interstellar regions where the temperature of neutral hydrogen will be no more than $100°K$ ($-173°C$) and may be as low as $50°K$, the hydrogen atoms will all be in their ground state – in other words, they will emit no visual or infra-red radiation, and except for their radio emissions are invisible. If it were not for the 21 cm emission the very existence of the hydrogen would be unknown.

The radio astronomer has been able to do more than merely plot the positions where neutral hydrogen is to be found, for this thermal emission is not always at precisely $21 \cdot 1$ cm; it may vary by a few tenths of a millimetre. A small part of this variation will be caused by a Doppler shift in wavelength owing to random motion of hydrogen atoms within the cloud, towards or away from the observer, but the major part will be caused by the motion of the cloud as a whole. Careful observations can therefore show up large-scale motions of cold hydrogen gas in various positions within the Galaxy, and this is just what van de Hulst, A. B. Muller and J. Oort at Leiden proceeded to do as soon as enough observations had been completed. Moreover, they took into account the fact that the Sun is revolving about the centre of the Galaxy, and making allowance for this, they then computed the rotational velocities of the clouds of neutral hydrogen, realizing that the radial velocity observed from the 21 cm emission would be due to, and part of, this rotational motion. They were also able to deduce the mass of hydrogen gas causing the emission by an analysis of its intensity.

They discovered that the total amount of neutral hydrogen appeared to lie between one or two per cent of the total mass of the Galaxy and, more important still, they found that the gas seemed to be distributed in long arms, spirally placed with respect to the centre of the Galaxy. Meanwhile Frank Kerr and James Hindman in Sydney were completing an analysis of 21 cm observations in the southern hemisphere. Their results led to a similar conclusion and it now appears that the Galaxy possesses two spiral arms, one some 16,000 light years from the centre and the other at about 21,000 light years. The spiral arms are thin, and seem to have a thickness not much greater than 700 light years, but they split up into a number of branches and even into separate strands.

This revelation of the existence of clouds of neutral hydrogen and, more especially, of the spiral arms of the Galaxy and even details of their structure, has been a notable contribution of the radio astronomer, since the density of optical material between the observer and more distant reaches of the plane of the Galaxy (in which the arms lie) has prevented the visual astronomer from observing the arms to any satisfactory degree. But the evidence obtained from 21 cm emission has not been the only factor that has brought new insight into the structure of the Galaxy; non-thermal emission has also played its part.

When radio observations are made close to the plane of the Galaxy, a continuous background of radiation is obtained over a wide range of wavelengths. On metre wavelengths the Milky Way (which visually marks the plane of the Galaxy) appears very wide, but on shorter radio wavelengths it appears more like a thin line, becoming progressively weaker as reception nears centimetre wavelengths. This narrow width of radiation seems to come from a disk some 1600 light years thick and is similar in this sense to what optical observations lead one to conclude. The radio radiation is of synchrotron origin. There is also a broad area of non-thermal radiation, and this is interpreted as coming from the halo of the Galaxy, composed of globular clusters, stars, and gas. Since it is a synchrotron emission, it must be associated with a magnetic field, which seems to be two or three times smaller than the field in the plane of the Galaxy. Bernard Mills has concluded that the halo is elliptical rather than spherical, stretching some 100,000 light years in the direction of the Galactic plane, and about 65,000 light years in a direction perpendicular to this. Other observations, made by C. A. Shain in Australia and Gart Westerhout in the United States, have made it clear that there is some absorption of non-thermal radiation, particularly when one comes close to the Galactic centre, this absorption being caused by ionized hydrogen. In brief, then, the general radio picture of the Galaxy gives neutral hydrogen along the spiral arms, some of which is ionized

VIB. (*left*) The elliptical galaxy NGC 4486 (M 87) photographed with the 200-inch telescope using a normal exposure. It shows no peculiar characteristics.

VIC. (*right*) The same galaxy photographed with a shorter exposure to show central details. The stimulus for the re-photographing came from radio observations and the jet has since been further investigated both optically and in the invisible wavelengths.

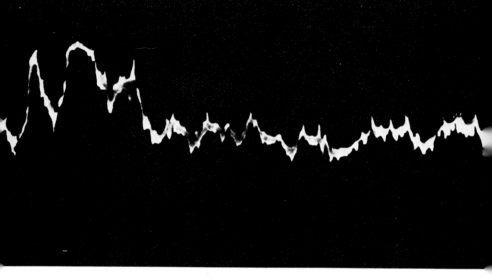

VIIA. A visual recording by Professor Drake using an oscilloscope of the pulse emitted by a pulsar. (The multiple pulse is shown on the left. The small irregular waves are caused by electronic 'noise').

VIIB. An array of cosmic ray recorders in the cosmic ray observatory at Ithaca, New York.

by radiation from hot bright stars, which are to be found in plenty in the arms. In addition there are high speed electrons and heavier (nuclear) particles that are circulating in the Galaxy, some in the halo and others confined to the spiral arms.

This is new and valuable knowledge, since what applies to our own Galaxy may be expected to apply to other spiral galaxies of a similar size, and it supplements what is available from visual observations. Equally is this so when it comes to studies concentrated on the centre of the Galaxy, for here optical observation can hardly penetrate at all; the centre of the Galaxy lies in the region of the dense Milky Way star clouds in Sagittarius, but much more than this is difficult to determine, as there is so much interstellar matter to absorb the visual radiation. At radio wavelengths, the interstellar matter has little absorbing effect. Observations made with radio telescopes show rather different emission areas at different wavelengths, as might be expected, just as visual observations through coloured filters give different pictures of the same object – a fact that, as already mentioned, has been put to good use by Fritz Zwicky. At 22 centimetres, the central regions of the Galaxy display a concentrated source with an apparent diameter of no more than $\frac{1}{2}°$, and this seems to be superimposed on a region of more extended but less intense emission, covering an area about 1° by 2°, but centred on the same point. At metre wavelengths, Mills has found that, using 3·5 metres, the centre still shows the extended emission that appears at 22 cm, but the small concentrated source vanishes, and, where it 'ought' to be, there is a minimum of radiation – in other words, at 3·5 metres, there is absorption in the very centre. Shain, using 15·2 metres, found the same kind of effect, but on this wavelength the absorption was even more pronounced. So far, then, it appears that there is an extended area of non-thermal radio radiation in the central regions, with some absorption at the centre due to ionized hydrogen.

However, it is not quite so simple as it appears, since Rodney Davies at Jodrell Bank believed that observations made with the 250-foot parabola showed evidence of a localized radio source in a nearly spiral arm, some 10,000 light years away. Again, F. D. Drake using a wavelength of 3·7 cm and obtaining a resolution of 6′ with the 85-foot parabola at West Bank, discovered that the strong central core observed by Westerhout was really four separate sources, one of which coincided with the centre point that Westerhout had determined and was the strongest. This Drake believed to be the nucleus of the Galaxy, and it began to seem as if the centre of our Galaxy was similar to that of the Andromeda galaxy, provided ionized hydrogen could be detected there.

Further investigation makes this appear likely, and studies of the central Galactic regions make it seem that, firstly, there is a small and intense central core, similar to that observed in the Andromeda galaxy. Secondly, from evidence helped by observations of a Doppler shift of the 21 cm line, there is a rotating mass of neutral hydrogen stretching out from the centre to a distance of some 20,000 light years, with a very strong magnetic field confined in a small central area. Thirdly, beyond this rotating mass of gas, the neutral hydrogen is expanding, being mainly contained in spiral arms, and may be replenished by material from the Galactic halo. Finally, closer in at 10,000 light years from the centre is evidence of a high hydrogen concentration, some ionized and some neutral, and this additional concentration appears to be associated with a spiral arm (figure 14).

These are important results, and it would not be too much of an exaggeration to say that, since the detailed radio astronomical studies of the Galaxy commenced some fifteen years ago, more has been discovered about its shape and construction than in the previous century and a half. And in one sense, this kind of investigation is only just beginning to yield results. Hydrogen emission at 5·99 cm has also been detected, while there has been found to be an important helium radio radiation at 12·3 cm; but full investigations at these wavelengths have yet to be completed. However, the hydroxyl (OH) molecule, composed of one oxygen and one hydrogen atom, has been found radiating at a wavelength of some 18 cm, and although studies of this radiation are far from complete, they have contributed to the picture of the Galaxy already sketched, and have made it clear that in interstellar gas, the OH molecule is being excited in some way not at present fully understood. Hydroxyl radiation is to be found at the edges of regions of ionized hydrogen gas, strongly polarized and with an unexpected distribution of energy over its very narrow wavelength range (18·6 cm to 17·4 cm); it also varies considerably in intensity with time. Again, observations at the National Radio Astronomy Observatory at Green Bank, West Virginia, have shown the existence of water, of ammonia molecules (NH_3) and formaldehyde (H_2CO); the last is an organic molecule and its presence affects the question of life in the universe.

But radio astronomy has not been confined to an analysis and investigation of the Galaxy alone. Inside and beyond it a number of discrete radio sources have been found, and, like other sources generated within the Galaxy, may be divided into those displaying thermal radiation, and those with non-thermal emission. The thermal sources are those associated with gaseous nebulae, in which radio radiation is generated

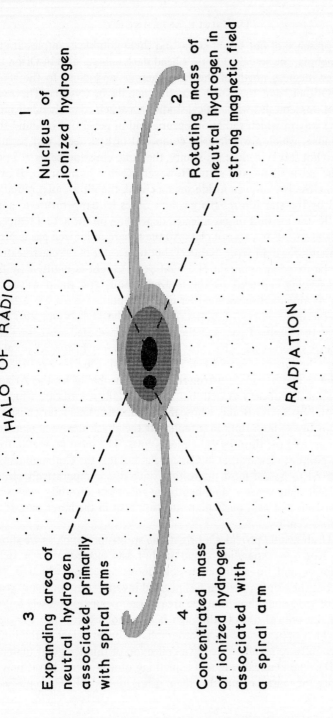

HALO OF RADIO

1
Nucleus of
ionized hydrogen

2
Rotating mass of
neutral hydrogen in
strong magnetic field

3
Expanding area of
neutral hydrogen
associated primarily
with spiral arms

4
Concentrated mass
of ionized hydrogen
associated with
a spiral arm

RADIATION

Figure 14

A sketch of the Galaxy, showing the elliptical shape of the radio 'halo' and the presence of neutral and ionized hydrogen as found from radio astronomical observations.

by the presence of hot bright stars, and three notable examples are the Orion nebula, the so-called horse's head dark nebula, also in Orion but some few degrees north of the large nebula and close to the star ζ (Zeta) Orionis, and the North America nebula in Cygnus. The radio observations, coupled with optical studies, have helped to elucidate more fully the way in which radiation occurs, and in general it is found that the nebulae, which are primarily composed of hydrogen, will be inert when no hot star is in close proximity, the radio emission being at 21 cm from the gas in its normal state. However, when a hot star (O, B or A class) is close by, or even inside such a cloud, the ultra-violet radiation emitted by the star ionizes the gas and raises its temperature to some 10,000°K, the amount of gas affected depending on the actual temperature of the star. For a star of class A0 the sphere of ionized gas is probably about three light years diameter, but with a star of O5 classification, it may be as great as about 500 light years. Light is emitted by free electrons being captured by the ionized atoms, and radio waves are emitted as already discussed earlier in the chapter.

Non-thermal sources of radiation within the Galaxy are also concerned with clouds of gas, but these are all associated with stars that were once supernovae. The most famous, from a historical point of view, are the supernovae of 1054, 1572 and 1604: the first was observed by Chinese and Japanese astronomers (and probably by the Arizona Indians and Europeans as well, but the important records are Chinese), the second by Tycho Brahe, whose measurements of planetary positions were to prove so important in formulating the physical laws of planetary behaviour, and the third by Johannes Kepler. The most powerful, from a radio point of view, is that in Cassiopeia, but it is the Crab nebula that has proved to be the most interesting since it is comparatively close – some 3580 light years – and has been an object of study ever since William Herschel paid particular attention to it in the latter part of the eighteenth century.

The Crab nebula provides a bright line spectrum – such as we should expect from an expanding cloud of hot gas ejected after a stellar explosion – and, in addition, a continuous spectrum which is very strong but the origin of which has puzzled investigators for many years. Radio observations show that the area of radio emission closely follows that of the optical object, although it does not extend into the outer parts which appear only on photographs taken in wavelengths longer than the infra-red, and are really a collection of thin irregular filaments (plate II). The filaments are still something of a mystery, but it may be that since the matter composing them is not ionized (the Crab possesses two small, visually faint stars), they are maintained in an active state by the powerful magnetic field within the main part of the nebula, and what may be a neutron star with a magnetic field has been detected there recently (see p. 110). The main radio emission comes from synchrotron radiation, an explanation first proposed in 1953 for this very object by Shklovsky and confirmed after optical and radio examination of the polarization of radiation from the nebula; synchrotron radiation is the cause of the continuous spectrum in its infra-red and visual wavelengths as well. Where the high speed electrons come from is uncertain: possibly they were derived from the supernova explosion, perhaps they are still being ejected from the central regions of the nebula or, a third alternative, they may be derived from a bombardment of the gas atoms by high velocity nuclear particles in space.

The source Cassiopeia A is rather different, since there is reason to believe that its supernova explosion produced a dimmer shell than in the case of the Crab or the 'historical' supernovae mentioned a moment ago, and a fluctuating emission of radiation – again different from these other supernovae. It does not seem to have been noticed historically, probably because of its dimmer shell and the absorption of radiation by interstellar dust and gas over the 11,000 light years that separate it from us. Nevertheless, Cassiopeia A is the strongest source of radio radiation in the sky, and J. Steinberg and J. Lequeux at Nancy in France have discovered that this radiation originates in an outer shell of gas only, and is decreasing by some one per cent per annum. Synchrotron radiation is the cause of the emission and it appears that although the original explosion was not so bright as in the case of many other supernovae, the mass of material involved was very large.

Perhaps the most dramatic, and certainly the most significant radio astronomical observations have been made of sources outside our Galaxy, the results of this work making it necessary to consider some important revisions of the general picture of the universe drawn from optical observations alone. Radio radiation is received both from the normal large galaxies known and observed for some years and also from other objects which are often referred to as 'radio galaxies', and although the latter are the more unusual and important from the point of view of the development of the universe, it will be more convenient to deal first with the normal galaxies. The radio radiation emitted by these objects is comparatively small and studies have been confined to those that do not lie too far away in space, such as the Andromeda galaxy, the Magellanic Clouds and other galaxies that are members of the 'local group'.

The Magellanic Clouds, which appear to be rotating about each other but may, as mentioned in Chapter 4, be part of a triple galactic system with our Galaxy as the third member, seem from radio obervations to be far richer in hydrogen than our own. On the other hand, the elliptical galaxies in the group give evidence of having considerably less, while the Andromeda galaxy, which, on present estimates of distance, is a rather larger spiral than ours, is similar in hydrogen content. The main emission from the Andromeda galaxy lies at a wavelength of 73 cm and is coincident with the centre of the galaxy, although its halo also radiates and from an area that is about two and a half times that computed for our Galaxy. In general, there seem to be no especially surprising features about the radio emission from normal galaxies, and the one important fact that has been derived from the observations is that the amount of hydrogen is proportional to their intrinsic luminosity. This provides no antithesis to the visual picture; elliptical galaxies have less hydrogen and spirals more, the quantity varying according to luminosity, but since bright population I stars are associated with hydrogen in the spiral arms, the radio observations lead to conclusions precisely similar to those derived from the analysis of the optical astronomer. It is the radio galaxy that produces surprises.

Radio galaxies are objects that are visually faint but strong emitters of radio radiation. Their dimness in visual wavelengths, and their apparent insignificance on photographs, led them to be largely ignored by the optical astronomer, whose attention was, naturally enough, concentrated on the brighter galaxies that could readily provide sufficient radiation for further analysis. It was only when the radio astronomer began to draw attention to certain patches in the sky, that visual examination was carried out in detail. To begin with there was another factor – the problem of resolution. The radio astronomer could recognize the existence of 'bright' radio sources, but he was unable to pinpoint them with any great precision; yet if the optical astronomer was to try to photograph the object, it was necessary for him to be able to know the location not to within 1' but to an accuracy of 2" or 1", otherwise his photograph would show too many objects, any of which could be the radio source. It was in 1954 that Walter Baade and Rudulph Minkowski made the first certain optical identification of a radio source, using the 200-inch reflector at Mount Palomar; to radio astronomers the object was known as Cygnus A and, optically, it lay in a rich field of stars.

Cygnus A is a double radio source with its radio components 1·8' apart, as determined by Bernard Mills and Graham Smith, and with an output of radio emission far greater than any other discrete radio source except for Cassiopeia A. Optically it is a dim object of magnitude 17·9, appearing like a double galaxy, with the two components no more than 2″ apart, so that the associated radio components lie one each side of the optical object. Its distance appears to be some 700 million light years, computed on the basis of Hubble's law that velocity of recession increases directly with distance, and it shows great gaseous activity, the spectrum displaying strong emission lines, broadened by motion of the gases causing them. Yet in spite of all the visual and radio evidence, there is some argument about precisely what is being observed here. Certainly there are two galaxies, but are they in collision, a close pair in orbit, or one galaxy splitting apart? The idea of an orbiting pair is not very likely since the objects appear to be young and vigorous and both components have the same red shift, whereas there would be likely to be slight differences if they were in mutual orbit. They could be colliding, although if many other such sources are discovered this is an unlikely explanation, since a collision might be expected to be a rare occurrence. The explanation that they are two young galaxies splitting apart is not open to these objections and, in the light of other observations now to be mentioned, appears the most likely explanation.

Other sources observed and identified early on in radio investigations of extra-galactic objects were a double galaxy in the Perseus cluster of galaxies, in which Baade and Minkowski found the components to have a difference in radial velocities of more than 1800 miles per second, a double radio galaxy in Centaurus, and a strange elliptical galaxy in Virgo. Baade and Minkowski interpreted the two-component object in Perseus as a galactic collision, but this is not necessarily so – the radio activity could be caused by internal eruptions in one of the two galaxies. The double galaxy in Centaurus is a powerful radio emitter, again with two separate radio sources: visually it appears like a large spherical galaxy split through the centre by a broad dark band of obscuring matter. Its optical diameter is 20′ and, at a computed distance of thirteen million light years has a diameter of three million light years with a separation either side of the dark band of some 25,000 light years. The radio picture, however, shows an area that is 10° across in a direction perpendicular to the dark obscuring belt. Another radio galaxy in Fornax also presents a similar appearance.

The radio source in Virgo gives a visual appearance of a class E0 elliptical galaxy, 5′ across, but its radio diameter is twice this with a 36″ central area. The most unusual and interesting feature of the object is a bright blue jet, some 20″ long, a feature that was missed in early photographs since it was obscured by surrounding material from the galaxy and showed up only after radio observations had led optical astronomers to make shorter exposures in the hope of bringing out the central regions. In radio and visual wavelengths the jet seems to have two components and emit polarized waves. The Virgo galaxy – catalogued as M 87* – is not the only extra-galactic object to display internal eruptions. The irregular galaxy M 82 in Ursa Major is another interesting eruptive source, which emits radio waves from its central regions although not at very great strength. On optical wavelengths, Allan Sandage and Clarence Lynds, using the 200-inch telescope, have found it to possess filaments of fast moving gas travelling away from the main body of the galaxy in a perpendicular direction. With observed velocities of the order of 620 miles per second, the gas is travelling quickly enough to escape from the galaxy's gravitational field. And in addition to M 82 and M 87, there are also the Seyfert galaxies – small galaxies with bright centres that are very active internally, with movements of gases also great enough to escape from the galactic gravitational field. Discovered by Carl Seyfert, these galaxies are visually dim but have strong emission lines broadened because of the gas velocities; they emit at radio as well as visual wavelengths.

It seems, then, reasonably clear that many, if not all, radio galaxies are extra-galactic objects undergoing vast internal upheavals, often with the ejection of gaseous material and, presumably, with the ejection of nuclear particles as well. Since multiple galaxies are known to exist, it would seem more likely that sources like Cygnus A, the double galaxy in Centaurus, and so on are instances where multiple galaxies are perhaps forming or possibly being driven apart rather than being driven into collision. At all events, we have here some important phenomena that are already causing revision in ideas of galactic formation and making it obvious that more large-scale change occurs in the universe than had previously been supposed. But if the discovery and observation of radio galaxies has modified the outlook on galaxies in general, a possible revolution in outlook arose from the optical identification of a radio source known as 3C 48† in 1960. Cooperation between T. Matthews and J. Bolton, two radio astronomers from the California Institute of Technology, and Jesse Greenstein, Guido Munch and Allen Sandage of Mount Palomar, had led to this source being found: it appeared as a

---

* The M catalogue numbers are those given in the late eighteenth century by Charles Messier, who compiled a list of hazy objects that he found likely to be confused with telescopic comets, which present a somewhat similar appearance when approaching the Sun but are not close enough to produce a tail.

† Again, a catalogue number, denoting item 48 in the third sky survey of radio sources made at the Mullard Radio Astronomy Observatory at Cambridge.

blue star of apparent magnitude no more than sixteen, although it possessed a strong ultra-violet emission. The object also displayed emission lines which could not be identified. Strangely enough, it looked visually just like a hot star, yet its radio intensity was sufficient to make it clear that it could be no ordinary star. Soon after, two other objects, 3C 196 and 3C 286, were found to be similar star-like bodies, and when Maarten Schmidt took spectrograms with the 200-inch, the former appeared to have no emission lines (these were in fact weak and only observed later), while 3C 286 showed one broad emission line in the blue-green region of the spectrum. Yet another object, this time of apparent magnitude eighteen and catalogued as 3C 147, was found with an emission line in the yellow. It seemed, therefore, that some peculiar kinds of star did emit radio waves strongly enough to be received on Earth, although their apparent brightness was very low and led to the supposition that they were a long way off. For some while, these 'radio stars' remained a mystery, but something of their astounding nature burst upon the astronomical world in 1962 with the identification of a fifth object, 3C 273. Cyril Hazard, using the large 210-foot radio parabola at the Australian radio astronomy observatory at Parkes, near Sydney, used an occultation of the source by the Moon to determine its position accurately. His method was simple and ingenious: he decided to observe the source, whose position was obtained from the third Cambridge catalogue, as it became obscured by the Moon in its orbital motion round the Earth, and to watch it again as the Moon moved away from its position. Since the Moon's position and orbital movements are known with considerable accuracy, Hazard believed he should be able to re-determine the position to within a second of arc at least. His observational technique was highly successful but the source turned out to be a double one, with the two components 20″ apart.

Clearly, this was reminiscent of a radio galaxy rather than a star, and the mystery became no clearer when the spectrum was examined, for John Oke found a broad emission line in the infra-red, using the 200-inch telescope, while Maarten Schmidt observed a line in the ultra-violet as well as other lines, yet none could be identified with any known substance. Fortunately, however, while Schmidt was reviewing his spectrum observations, he suddenly realized that if the ultra-violet and infra-red lines were neglected, the others had a familiar pattern: the reason they had not been identified before being that they all displayed an unprecedented red shift. The infra-red line then turned out to be a line due to doubly ionized oxygen, a line which is normally associated with gaseous nebulae inside our Galaxy; the ultra-violet line was caused by ionized magnesium, normally so far in the ultra-violet that it is invisible from Earth-based observatories. Obviously this was no star, whatever else it might be, and the name quasi-stellar-radio-object was used to describe it, a term that has become shortened to quasar.

Other analyses showed that 3C 48's spectrum could likewise be explained by a red shift, but in this case a shift of thirty-seven per cent, or over twice that of 3C 273. There immediately arose the problem of explaining such large shifts, and two possible answers presented themselves; one that the shifts were due to a prodigious velocity of recession, greater than any previously observed; secondly that they were due primarily to an Einstein shift – a red shift caused by the gravitational field of the body. The latter explanation was soon disposed of, since, if the two quasars had their immense red shifts caused by their gravitational fields, they must be surprisingly dense small objects – if 3C 48, for instance, had a mass equal to that of the Sun, and was no more than six miles in diameter, it could give the observed red-shift – but in order to present the hydrogen line as intensely as it did, its distance could then be no more than nine miles, an obviously irrational answer. Yet even if it were supposed that the quasar had a mass some 10^5 times that of the Sun, it would still have to lie within the solar system and so massive a body would have made its existence self-evident from its effect on planetary motions long before its discovery by radio observations. There seems, therefore, no alternative but to interpret much of the red shift as evidence of a high velocity recession, and this has been confirmed by subsequent observation where, in one case, a quasar has been found with a red shift of 212 per cent, a figure that would be impossible to explain on an ordinary gravitational hypothesis.

The red shift is a velocity shift, but this still permits of two possibilities; one that the quasars are galactic phenomena – the débris perhaps of a gigantic explosion within the nucleus of our Galaxy; secondly that they are extra-galactic sources lying at very great distances. The galactic explosion hypothesis holds some attraction, especially as eruptive processes within galaxies have been observed to be taking place, but it possesses one great drawback: it offers no explanation why all the quasar spectral line shifts are red and none blue, yet if the quasars are caused by a galactic explosion one would expect as many to be moving towards us as away from us, and certainly not expect there to be only movements of recession. Possibly, if the quasars were part of a jet of ejected material this would not be so unlikely, but on the other hand they seem separate compact bodies and there is no observational evidence of the kind to be expected if a tail of ejecta occurs in our

Galaxy. Greenstein and Schmidt, certainly, have concluded that the quasar shifts are velocity shifts of extra-galactic objects and, on the hypothesis of the expanding universe, are able to compute distances, since, the further off an object lies, the faster it must be moving. This gives very large distances for quasars, of the order of thousands of millions of light years, making them the most distant objects so far observed: indeed, in the case of quasar 1116 + 12 the estimate made from the shift of its emission lines gives 8000 million light years. Yet surprising though this is, an investigation by Geoffrey Burbidge on the different red shifts for every quasar given by shift of absorption lines compared with emission lines makes it seem likely that there is considerable obscuring matter between quasars and ourselves, and to such a degree that great distance may be the only logical explanation.

At such large distances, some abnormal conditions must exist if the body is to generate the amount of energy which it is observed to radiate, and one of the great problems facing astrophysicists is to account for such generation by modern physical theory, otherwise a drastic revision of some of the concepts of basic physics may be necessary. A number of suggestions have been made to account for their output of at least some ten times that of a normal galaxy. Geoffrey Burbidge suggested a chain of supernova explosions, triggered off by one such explosion in a body of small dimensions, and Thomas Gold, S. Ulam and Lodewijk Woltjer proposed a series of collapses and collisions within a dense star system, for it is clear that quasars occupy only a small volume. This is known not only from radio observations utilizing widely spaced interferometers, but also from the fact that some quasars at least vary their emission of radiation and do so in a period of not less than a year and no longer than about ten years – 3C 273 varies seventeen per cent per annum at 3 cm wavelength. This makes it necessary for their diameters to lie between one and ten light years, otherwise the variation would be unobservable, since light from the rear of the body would take longer than the variation period to travel from back to front; in consequence while radiation from the front would be at one intensity, radiation from the rear would be of different intensity and the variation would be masked.

There is, it has been suggested by Fred Hoyle and others, a different mechanism to account for quasar radiation, and that is an implosion instead of an explosion as envisaged by Burbidge. The implosion would have to be catastrophic and of the nature of a gravitational collapse, in which the whole body contracts at a rate that makes even a neutron star seem slow by comparison. Such a collapse would release energy, but it is a complex process since it involves relativity considerations, and it will be better to postpone a fuller description until Chapter 10. However, if gravitational collapse is the mechanism by which radiation is emitted at so high a rate, it is necessary that the original body must have possessed a mass 10^8 times that of the Sun – a very high figure.

Greenstein and Schmidt have constructed a slightly more conventional model, the centre of which is a very small extremely dense object that generates most of the radiation and is surrounded by ionized gas at a temperature of 15,000°K, with a great density and a mass equivalent to about one million Suns. The emission lines would be generated, they suggest, within the gas cloud, the diameter of which is of the order of one to ten light years. Such an object would be likely to possess a very strong magnetic field in which synchrotron radiation could be generated at the rate required. Still other proposals have been made, H. C. Arp of Mount Wilson and Palomar observatories suggesting that quasars may be ejected from the more unusual kind of radio galaxy, in which case the Doppler shifts may not, he believes, be interpreted wholly on the expansion hypothesis, and this would bring them closer. But, again, we should expect some to have blue shifts or very small red shifts at the least if the ejection takes place in all directions, as one would imagine. Moreover, an absorption line at 21 cm in the spectrum of 3C 273 makes it seem likely that the absorption is due to the Virgo cluster of galaxies, which lies at some thirty million light years and, if it is, can only be explicable on the supposition that this quasar is further off; but what is true for 3C 273 is likely to be true of the other 200 quasars observed.

More novel proposals have come from J. Barnothy, H. Alfvén and W. Teller, Barnothy suggesting that the quasar is a 'radiation illusion', having an apparently small size in visual and radio wavelengths only because its radiation is concentrated by a Seyfert galaxy lying directly in the line of sight of a larger galaxy, the Seyfert galaxy acting like a gravitational lens. This suggestion is generally thought to be too improbable in view of the number of quasars observed, since the likelihood of so many Seyfert galaxies lying precisely in the line of sight seems to be stretching the arm of coincidence a little too far. Alfvén and Teller have taken a different approach, and their suggestion is that the immense release of energy observed is due to the annihilation of particles of anti-matter with those of ordinary matter. The problem here is where and how anti-matter is formed in considerable quantity; but it is conceivable that galaxies of each kind of matter do condense in separate areas and that possibly material ejected from one kind of galaxy meets material from or within another.

Whatever the answer to the mechanism of energy generation of quasars may be, it is clear that it is not possible to come to any firm decision at the moment. Ryle, however, has pointed out that recent mapping of sources with the one mile aperture synthesis radio telescope at Cambridge indicates that 3C 47, for instance, presents two separate sources that are very reminiscent of a radio galaxy. Moreover, there is a marked cut off at a particular wavelength in the radio radiation from quasars, and when this is used as a basis for computing the total energy output the answer again comes close to that of a radio galaxy. Ryle has also drawn attention to the fact that at high frequencies the radio emission from a Seyfert galaxy like Perseus A varies in a similar way to that of a quasar, and it is clearly his opinion that the three classes of object must be closely related.

The quasar has opened a new chapter in astronomy and is bound to have an effect – probably of great consequence – on theories of the birth or death of galaxies, and therefore on problems of the behaviour of the universe as a whole. But quasars are not the only significant contribution of radio astronomy to cosmology, for Robert Dicke has analysed what appears to be a background radio radiation that could be due to an earlier condensed state of the universe. The details and impact of this will, however, fit more properly into the discussion of Chapter 10.

Finally, there is still another new phenomenon which the radio astronomer has discovered – the pulsating star. A theory of pulsation has for some time been derived to account for the behaviour of many kinds of variable star, but the radio astronomers' discovery seems to refer to a different kind of object, and has been given the name 'pulsar'. First detected in August 1967 at the Mullard Radio Astronomy Observatory at Cambridge University by Jocelyn Bell and Anthony Hewish, they have been the subject of widespread study. The essential quality of the pulsar is that it emits regular and very brief pulses of radio radiation at frequent intervals – typically a pulse that lasts no more than about forty milliseconds (forty thousandths of a second) and which is repeated every one and a third seconds or less – at least this is so for three of the first four sources discovered, while the fourth emits a shorter pulse – of only eighteen milliseconds – and at a shorter interval, namely every quarter second. The pulses themselves may display one, two or three peaks at each appearance. The regularity of the emission is remarkable.

Investigations of the pulsars so far known make it evident that they lie within our Galaxy, probably within 850 light years and inside, or near to, a local spiral arm. After a first humorous description of these sources as LGM's ('Little Green Men') on the basis of a very tentative and not very serious suggestion that the observations were of messages from intelligent beings elsewhere in the universe, careful analysis and optical observations have led to a number of possible physical explanations. One, to account for the very short duration of the pulses, and suggested by J. S. Thorne of Chicago University and J. S. Ipser of the California Institute of Technology, is that what radio astronomers are observing are the higher harmonies of a more fundamental oscillation that some types of star undergo. But since this idea was formulated, new experimental evidence has become available. A slowing down of pulse rate has been observed – this amounts to about 4 seconds per year on the average – and visual pulses have been detected (1969) in the pulsar NP 0532 by W. John Cocke and his colleagues at Stewart Observatory, and photographed at Lick Observatory. This pulsar lies in the Crab nebula, where there have been suspicions of the existence of a neutron star, and this may be what a pulsar is. T. Gold has suggested that a pulsar is, in fact, a fast rotating neutron star with an extended 'bar' of ionized gas projecting from two sides. The size of the neutron star is calculated as some 10 miles in diameter and the ionized bar is thought of as extending some 100 miles each side. The magnetic field of the fast rotating star would sweep the bar round with it and this would cause radiation emission from the ends of the bar in a way analogous to synchrotron emission. This could account both for pulses at radio as well as visual wavelengths. Another possibility is that a pulsar is a very close white dwarf or neutron star binary system, rotating rapidly.

Whatever the final answer may turn out to be, it is clear that these observations are causing a thorough theoretical investigation of the precise way neutron stars behave. They are bound, therefore, to exert a profound effect on our knowledge of stars in the later stages of their existence.

CHAPTER 7

Cosmic Ray and Particle Astronomy

The discovery of radioactivity by the Curies in the late 1890s awakened considerable interest in the electrification or ionization of gases (i.e. the removal of electrons), and in the electrical charges associated with the emission of radiation, so that many physicists turned their attention to the problems that arose. At the Cavendish Laboratory at Cambridge, much investigation into radioactivity and atomic particles began, and it was here that in 1911 C. T. R. Wilson developed his 'cloud' chamber – essentially a metal cylinder with a glass top in which atomic particles could make their presence known by their collisions with particles of very moist air, ionization taking place along the path of the atomic particle and water droplets forming on the ionized water vapour molecules of the air in the cylinder. The ionizing effect of radioactivity on gas in an evacuated tube was well known, as was its power to ionize air molecules and cause an electrometer* to become discharged, but as investigations continued it was discovered that even without the presence of a radioactive material, no delicate electrometer would remain permanently charged. The discharge could not be due to dampness, or to the presence of dust, for it occurred even if the electrometer were kept in dry clean air, and since the radioactive discharge was due to α-, β- or γ-radiations, Wilson suggested that it might be caused by some kind of extra-terrestrial radiation. In 1909 the physicist Göckl ascended in a balloon to discover whether the discharge altered at different heights above the ground, and others, notably Victor Hess and Kolhörster, ascended for the same purpose: in every case they found that the greater the height the stronger the discharge. These ascents were made before the First World War, and after it the investigations were continued with some surprising results. In 1922 R. A. Millikan and I. S. Bowen in the United States ascended to a height of more than ten miles and then, three years later, Millikan with his colleague Cameron lowered electroscopes to a depth of seventy feet in water that they checked was radium

* The electrometer can take a number of forms, but essentially it is a delicate device that indicates the presence of an electric charge by the attraction or repulsion of two metal plates, pieces of gold leaf, etc.

free, and noted a continuous decrease in the discharge rate as the electroscopes descended.

It was clear, therefore, that something was causing a continual electrical discharge by ionization of the air surrounding a body, and that such a
discharge varied with distance above or below the Earth's surface, and
there was every indication that this was due to some form of celestial or
cosmic radiation. Measurements showed that the radiation possessed
immense energy – of the order of 6 Gev (6×10^9 ev) – and further studies
made it clear that while γ-radiation was involved, the radiation was for
the most part composed of atomic particles. The reactions of the particles were investigated, evidence was found for the existence of the
positron, which, as already mentioned, Paul Dirac had concluded should
exist if his quantum theory about the emission of energy in discrete
amounts were correct, and a whole new field of study was initiated.

Over the years, and particularly since the end of the Second World
War, much attention has been devoted to the problem of cosmic radiation, and the general nature of the radiation and of the particles involved
has become clear. Two kinds have been recognized – primary and
secondary radiation – and although the secondary was discovered and
studied first, it will be more convenient to reverse the historical sequence
and consider the primary before the secondary. Primary cosmic radiation
consists of nuclear particles travelling with very high velocities and therefore possessing considerable energy. As soon as they arrive in the
terrestrial atmosphere, they react with the molecules and give rise to
secondary radiation: few primary particles reach the ground. Investigating primary radiation is not easy, since observation must be made
either at the upper limits of the atmosphere or out in space, but the
development in the last decade of automatic high-altitude balloons,
rockets such as the Aerobee or the Skylark, and Earth-orbiting space
probes, has rendered it possible to make some very detailed measurements. The instruments launched in balloons, rockets and probes for
detecting the high energy primary particles are of one of three main
kinds – special photographic plates, Geiger counters and scintillators.
The photographic plate method of detection has been used for many
years, notably by Cecil Powell at Bristol University, and in essence consists of a pile of photographic plates fastened tightly together. The plates
have photographic emulsions specially designed for the task of recording
the passage of nuclear particles through them, and when processed each
passage becomes visible through a microscope, as a black line. Invariably
this line breaks into branches, indicating the mass charge and velocity of
the particle, and its subsequent decay and transformation on striking the

VIII. The launch of the improved Skylark rocket, with new stabilizing devices. This simple and comparatively inexpensive rocket is excellent for work up to heights of about 100 miles above the ground.

IX. Checking Orbiting Solar Observatory (OSO-4) before launch at Cape Kennedy. The craft weighs 594 lbs (270 kilograms). The semicircular area is covered with solar cells to provide electrical energy from sunlight for OSO-4's equipment: the protruding arms carry devices for measuring radiative conditions away from the vehicle and the observatory also contains equipment for observing solar flares.

nucleus of an atom in the photographic emulsion. It is an admirable technique, since the plates can be examined and re-examined at leisure on their return to the ground, but it is obviously limited to use in equipment that can be recovered after launching – to balloons and to small rockets that fall back to Earth after reaching 100 miles or so.

In probes or artificial satellites in orbit round the Earth, it is not always practicable to recover observing equipment, and devices must be employed that can provide electrical impulses which may be transmitted back to Earth by appropriate radio equipment. The Geiger counter – or more correctly the Geiger-Müller counter, named after its inventors Hans Geiger and C. Müller – consists of a metal tube containing gases (usually argon and alcohol vapour), with a very thin wire electrically isolated from the tube but passing along its axis. The wire and the metal tube are electrically charged by connection to a battery (the tube negatively and the wire positively), and when a primary cosmic ray particle enters the chamber it will ionize the gas, and the electrons freed will then be attracted to the wire, some ionizing more gas atoms on the way, so that the number of electrons reaching the wire will be larger than the electrons freed by the original primary particle. The arrival of the electrons at the wire is equivalent to a short burst of electric current – an electrical current is, after all, a movement of electrons – and the impulse obtained may be recorded and later transmitted back to Earth. The scintillator is simply a device consisting of a crystal that emits radiation when struck by a high-speed nuclear particle – an action analogous to that of luminous paint in wrist-watches, in which radium is mixed with a substance such as zinc sulphide, the latter emitting photons of visible wavelength when struck by the helium nuclei (α-particles) ejected from the radium. The light emitted by the scintillator, which may be of very low intensity, is usually passed to a photomultiplier tube to amplify the energy from the pulse of light to enable a recording to be made, the photomultiplier consisting of an evacuated tube in which there are a number of metal plates coated with a light-sensitive substance (caesium, for example). The scintillation falls on the first plate of the photomultiplier and electrons are emitted from it and pass to the next plate, which then emits an increased number of electrons, and so on down the tube: the amplification of a current that may be obtained in this way is of the order of a million and thus permits the detection of extremely dim scintillations.

In the terrestrial laboratory, which a few primary particles may reach and where secondary cosmic ray emission is to be found, additional equipment may be used. The cloud chamber has already been mentioned, and one of its advantages is that it may be placed within the field of a large electromagnet so that the electrical charges on particles may be unambiguously determined, but it is not the only such device now in use, indeed it possesses a number of disadvantages that render some improvements necessary. The greatest drawback of the cloud chamber is that the trail of ionized damp air particles persists for a comparatively long time so that the number of high-speed particles that may be admitted must be limited if the tracks are not to become too confused for analysis. To try to overcome this, Donald Glaser of the University of Michigan developed a new kind of detector that has been in wide use since the late 1950s and which, incidentally, earned him a Nobel prize. The detector, known as a bubble chamber, is a cylinder containing a liquid that is close to its boiling point. By the partial withdrawal of a plunger, the liquid expands and then reaches a density at which it ought to boil (since any liquid boils at lower temperatures, the lower the pressure under which it is kept). This 'super-heated' liquid takes time to boil – there is, in fact, a delay while the bubbles that are generated by the act of boiling occur – and if, during this delay, nuclear particles are admitted to the chamber, their tracks will become visible since bubbles will form along them and do so more quickly than they form in the remainder of the liquid. The active period lasts for a very short time, and in practice the plunger is operated at frequent intervals – it takes the chamber something like a second to cool between one superheating and the next. The tracks in it are very thin and clear, but since it cannot be set off only when nuclear particles appear, but operates every time the plunger is withdrawn, a photograph must be taken at every superheating: this entails the examination of thousands of pictures to detect the reactions. In contrast, the cloud chamber can be put into operation only when nuclear particles are detected (by a Geiger counter, for instance). However the bubble chamber has the advantage that a variety of liquids may be selected depending on the kind of nuclear particles expected: liquid hydrogen (boiling temperature $-257\,°C$), which presents the incoming nuclei with only protons and electrons, is frequently used, but if the presence of neutrons is likely to raise no problems, propane is employed (boiling point $-45\,°C$), or if high stopping power for nuclear particles is required then recourse is made to a heavy element like the liquified gas xenon (boiling point $-106\cdot9\,°C$).

A further development of the bubble chamber is the spark chamber, which became a practical proposition in 1959 due primarily to the work of S. Fukui and S. Miyamoto of Osaka University, Japan. It consists of a chamber filled with neon gas in which there are a number of parallel metal plates fed with a high voltage that is, nevertheless, not high enough to cause a spark discharge in the neon gas between one plate and the next. When a nuclear particle enters the chamber, the gas is ionized along its track (and along the tracks of the particles ejected by the reactions it causes); if a high voltage is then passed to the plates, a spark will travel between those plates through which the particle has travelled; the discharge also occurs only at those points between one plate and the next where the particle has travelled and the gas is ionized. The tracks of the particle and the result of its reactions may be seen and photographed, but since the ionization caused in the gas lasts for a short time, chambers are fitted with detectors that sense not only the arrival of a nuclear particle but also the reactions that occur, so that a photograph is taken only when the specific results required actually happen. The detectors have to operate quickly and, in general, the photograph is taken only 100 nanoseconds* after the arrival of the particle and the occurrence of its reactions, and in order to obviate confusion between one set of reactions and another, the plates are discharged and charged again once every few millionths of a second. The plates themselves may be made of metal foil, or of carbon (if reactions with carbon are required), or may even be hollow and filled with helium. Like the cloud chamber, the bubble chamber and spark chamber may be – and usually are – operated in a magnetic field. The advantages of the spark chamber are that the decision to photograph an event may be taken automatically and after the event has occurred – thus avoiding the examination of an unwieldy bulk of photographs – and that since the chamber is swept clean of ionization trails so frequently the pictures will show only the reaction being investigated.

More recently (1967) another technique designed to observe the arrival of primary cosmic ray particles with extremely high energies of the order of ten million Tev† (10^{19} ev) has been developed, and at Ithaca, New York, there is an example of this type of cosmic ray observatory. Such ultra-high-energy particles are rare – on the average one will appear over one square mile every twelve days – but when it does arrive it will react with the air and produce a shower of secondary particles which, when they pass close to nitrogen molecules, will cause the latter to emit visible radiation. This radiation will be dim and transitory, lasting no more than forty millionths of a second and probably far less. The detection equipment at Ithaca is extremely sensitive and it is claimed that it can detect – by observing this radiation – the

* One nanosecond is 10^{-9} or one billionth of a second.
† One Tev or tera-electron volt is one million million (10^{12}) electron volts.

arrival of one proton at a distance of nineteen miles; certainly it has a wide field of view that covers the entire sky, and it must be realized that since the nitrogen emission occurs in all directions, one may expect to observe most reactions wherever they happen. The equipment is composed of multiple units of photomultipliers, placed behind sixteen Fresnel lenses,* each of eighteen inches diameter; they are mounted in a small building, the roof of which has sixteen flat sections, each facing a separate area of sky. In front of each Fresnel lens is a filter which limits the incoming radiation to wavelengths between 3100 Å (ultraviolet) to 4100 Å (blue-green). Below the roof is a display of cathode ray tubes, one for each photomultiplier, and a camera. When a very high energy particle is detected it causes the secondary particles, and the tracks of these are rendered visible by the glow of the nitrogen molecules which is picked up by the photomultipliers; the cathode ray tubes depict the particle track, only those tubes operating for those photomultipliers that have detected emission, and a recording is made.

The primary particles that enter the terrestrial atmosphere are all moving with high energies, and are either protons or the nuclei of heavier atoms. The energies are not much less than 0·1 to 10 Gev and may reach as much as 10^6 Tev, as already mentioned. Those in the 0·1 to 10 Gev range are primarily, if not entirely, protons, and are strongly affected by the Sun's magnetic field. This effect is greatest at sunspot maximum when the so-called 'solar wind' (which will be described shortly) generates a kind of magnetic shield that prevents many particles from reaching the Earth: at sunspot minimum the shield has its least powerful effect and the influx of particles within this energy range is at its maximum. At higher energies, from 10 Gev to 10 Tev, the particles are composed of many heavier nuclei than those of hydrogen (protons), and nuclei of lithium, beryllium and boron are plentiful. Because of their higher energies they are less affected by the magnetic field of the Sun, but it is only when energies greater than 10 Tev are considered that the effect of the Sun or of any other magnetic field is negligible. Such ultra-high velocity particles although rare are nevertheless important; since they are virtually unaffected by magnetic fields, they may be considered as having travelled in a straight line from their place of origin

---

* A Fresnel lens is one in which the surface is cut in stepped concentric zones, each zone differing slightly from the next in the angle of its step or, to put it another way, in its radius of curvature measured from the focus of the lens. It is an inexpensive yet effective method for obtaining a good focus over a large area and is now much used as a viewing screen in miniature cameras. It was invented in 1822 by Augustin Fresnel in order to provide a clear parallel beam from lighthouses.

to the Earth: if, then, one can determine their direction, it may be possible to find their source.

Secondary cosmic rays are those caused by the impact of primary particles on the gas molecules of the terrestrial atmosphere, and a study of them is important because it can provide evidence of the mass and velocity of the primaries that have caused the observed reactions. The secondary particles may be observed by techniques already mentioned but, in recent years, John Jelley at Jodrell Bank has used radio telescopes to detect radio wavelength emission generated when a cascade of secondary cosmic ray particles appears. This work was initiated in 1965 and, although no precise method has been agreed for the generation of the radio frequencies observed, it has been suggested that possibly the positive and negative particles are separated by the Earth's magnetic field as soon as they are formed and, if this is so, the effect would be similar to a dipole moving rapidly and this, certainly, would generate a radio frequency emission. Also, since the secondary particles cause some ionization, it should theoretically be possible to detect this ionization with radar, for radio pulses would be reflected as they are from the ionized particles in a meteor trail. The radar method is being tried at Tokyo University by K. Suga, who hopes to be able to cover an area of a few hundred square miles over the Earth's surface.

The study of secondary cosmic radiation has provided the nuclear physicist with a vast amount of information, not least the first evidence for the existence of some of the mesons. For consider the effect of protons with high velocity entering the Earth's atmosphere: the protons will strike the nuclei of the oxygen and nitrogen atoms and will give rise to mesons. These are short-lived, and the neutral π-mesons will become annihilated on passing through matter, being transformed into γ-radiation. The negatively charged π-mesons, on the other hand, will be attracted to any nuclei through which they pass, and they will cause disintegrations of these nuclei, often breaking them up completely. However, most negative π-mesons, as well as those with a positive charge, will continue moving and then decay before meeting any other particle; μ-mesons are then formed, as well as neutrinos. The μ-meson does not interact with matter and these particles will therefore travel a long way before they too decay into an electron and two neutrinos.

The electrons generated by a primary cosmic particle will also have considerable energy, and a high energy electron, when slowed down by passing close to a nucleus, will emit radiation, and this behaviour is most marked at energies greater than 1 Mev. The radiation emitted will lie at X-ray wavelengths but in the atmosphere it is soon absorbed and gives rise to what is known as 'pair production' – it is transformed into a pair of positrons and electrons. The original electron continues, having emitted X-radiation, although its velocity is now reduced but not necessarily below 1 Mev, and it may well continue to emit X-radiation at subsequent decelerations when it passes the nuclei of other atoms until, in the end, its velocity will be too low to do so. But on each occasion when X-radiation is emitted, pair production occurs, and there are more high-speed electrons to be decelerated, emit X-radiation themselves, and so on. The whole procedure is multiplied and the original high speed electron can be described as initiating a cascade of particles. Such cascades are no mere theoretical process either, for they have been frequently observed.

But if secondary cosmic radiation and the analysis of primary cosmic ray particles is important to the nuclear physicist, they are also important to the astronomer whose main problem is to discover their source and, in doing so, obtain new evidence about the conditions and behaviour of different celestial objects. Because the Sun is at times in a comparatively active condition, it is an obvious first place to examine as a generator of high-speed nuclear particles, although since there appears to be no marked change in the influx of cosmic rays by day or night, it might be conjectured that the investigation is likely to be doomed to failure. However we are dealing here with electrified particles and thus with a corpuscular radiation that is affected by magnetic fields, and, as the Earth possesses a magnetic field, it is to be expected that this may have some effect on the behaviour of particles that reach our planet.

The occurrence of a glow in the night sky in the north and south polar regions – the aurorae – has been known for centuries and it was Edmond Halley who, as far back as the last decade of the seventeenth century, suggested an explanation that involved terrestrial magnetism, but it was not until 1929 that Sydney Chapman and Vincento Ferraro suggested that the aurorae and terrestrial magnetic storms were caused by a plasma emanating from the Sun. They opined that this plasma would sweep across the Earth's magnetic field and initiate all kinds of effects, some of which, like magnetic storms, were felt at ground level. Measurements made outside the Earth by artificial satellites and, in particular in 1962 by the space probe Mariner II and four years earlier by the probes Pioneer I and III and the satellites Explorer I, III and IV, have clarified the matter considerably, and it is now known that there is a streaming ionized gas that comes from the Sun. This may be thought of as pushing the lines of magnetic force closer to the Earth on its sunward side, and away outwards on the other, the night side. The solar wind, as this plasma is usually called, is now established but so, too, are two wide curved 'radiation' belts that surround the Earth. These Van Allen belts, named after James Van Allen who discovered their existence from the space probe investigations, are shaped rather like two ring-doughnuts; the Earth is in the centre with the first lying some 2000 miles from the ground and the other 10,000 miles. The belts were found to exist from an examination of the counts of nuclear particles obtained by Geiger counters in the space probes, and they are in reality areas in which electrons and protons (and possibly a few α-particles) are moving to and fro at considerable velocity. Their motions appear to be two-fold: on the one hand they spiral back and forth from pole to pole of the Earth along lines that are shaped like the lines of terrestrial magnetic force, and, on the other, they drift round the Earth. The direction of drift depends on the charge on the particles, the electrons moving in an eastward direction, and the protons westwards.

The outer belt seems to be associated with protons and electrons ejected from the Sun. Some of this material may well lie in the magnetic field around the Earth and drift into the outer belt, but on the occasion of solar flares, the protons and electrons arrive with velocities of some 600 miles per second – an energy for protons of only 5000 ev – but even these are trapped in the outer belt. The emission of such particles is not regular and the outer belt has been found to change considerably over a period of weeks and months. The inner belt, it is thought, is composed of secondary cosmic ray protons and electrons ejected upwards rather than downwards in the disintegrations caused by the arrival of primary particles. The belts can only trap particles with energies that are not too great, and are quite unable to capture primary cosmic ray particles. Yet it is now known that some very high speed and high energy primary particles are, in fact, ejected from large solar flares; for instance, during a large solar flare in February 1956, primaries with energies between 1 and 30 Gev were emitted. The Van Allen belts therefore show a connection with cosmic ray secondaries and an emission of particles from the Sun, but their very existence may be taken as evidence of the solar emission of comparatively low energy particles. Flares do, however, produce cosmic ray primaries, but the frequency of flares and the energy of the primary particles that they emit make it clear that the Sun cannot be the source of most of the primary particles that are received. It can in no way provide these particles with energies in the 100 Gev range or higher, nor are there sufficient flares, particularly at times of sunspot minimum, to account for the actual influx observed. A search must be made elsewhere in space for the majority of cosmic ray primaries.

Perhaps it is only to be expected, if solar flares are the source of cosmic ray primaries, that flare stars, mentioned in the last chapter, are perhaps a significant source. Unfortunately it is, at present, impossible to be certain whether or not flare stars eject particles with sufficient velocities, since techniques for determining the directions of primary sources are of poor resolution (due to the interference of the solar wind and the terrestrial magnetic field and Van Allen belts) unless the energy is many Tev. Theoretically they could act in this way, although the velocities of the primary particles that they could be expected to eject would not be ultra-high velocity and might lie in an intermediate range between the solar particles and the most energetic known. Possibly the use of orbiting space probes with appropriate directional counters, might decide the question, for it seems certain that no terrestrial observations can do so.

Some clues about the generation of primary particles within our own Galaxy have, however, been obtained in recent years, and in particular investigations on the number of particles that are continually arriving and the energies they possess have produced some interesting if perplexing results. For instance, the lower their energy, the greater the number of particles that enters the atmosphere – if their energy is only of the order of 10 Gev, observations show that on the average some 1500 particles arrive per second over every square yard of the Earth's surface, but at an energy of 1000 Gev (1 Tev) only three particles enter every ten seconds. The number becomes progressively less as still higher energies are reached and at 10^4 Tev the influx has dropped to three every eleven and a half days: here there appears to be a break in the sequence, and no higher energy particles are found until one reaches a value of 100 times – in other words energies of 10^6 Tev. Only then can figures for arrival rate be found: at 10^6 Tev, the average number is three particles every square mile (instead of square yard) each three days, dropping still further at higher energies. The presence of this energy gap must be explained, and it has been suggested that there is an upper energy limit for particles generated within the Galaxy. Should they exceed the limiting value – 10^4 Tev for protons, 2×10^4 Tev for helium nuclei (α-particles) which are twice as heavy, and 26×10^4 Tev for iron nuclei – then the particles would leave the Galaxy and travel into inter-galactic space. Those particles with energies greater than these may, on this interpretation, be assumed to come from outside our Galaxy.

The higher energy cosmic ray primaries that do arrive from interstellar space and are generated within the Galaxy, may be expected to come not from flare stars, but either from supernovae remnants or, very possibly, interstellar clouds of gas. The Crab nebula is, on this argument, a very likely source and is being examined with this possibility in mind. But it seems that the general opinion is that primary particles receive their high velocities from the magnetic fields within the Galaxy, since these fields exert considerable influence on the behaviour of its gaseous matter. The mechanism here would be analogous to that in a supernova such as the Crab, where magnetic fields are known to exist and synchrotron radiation is observed, and is known as spallation – the disruption of atomic nuclei with the subsequent production of unstable nuclei in their place, nuclei that will themselves decay. The energies to cause spallation are, it is thought, obtained from nuclear particles ejected from stars close to or inside the gaseous clouds, which are then accelerated as they come under the influence of the magnetic fields. Some of the primary cosmic ray particles thus formed may reach sufficient velocities to escape from the Galaxy altogether, others will not be ejected and accelerated to such a degree and will travel within the Galaxy, reaching the Earth or some other similar body. However it may well be that some become accelerated further by another magnetic field into which they venture, and then leave the Galaxy.

If cosmic ray particles are leaving our own Galaxy, then it must be assumed that they are ejected from other galaxies – at least from those of the spiral type in which gas clouds and magnetic fields are known to exist. Such particles will have high velocities and energies, otherwise they could not escape in the first place, and such high velocities enable them to pass directly through the solar wind and the terrestrial magnetic field, so that their direction is a guide to their source. Even so, no sources have been identified at present, although the use of such cosmic ray observatories as the one at Ithaca may be able to bring some evidence to bear, while space probes will doubtless be able to extract additional information. In particular it will be important to discover how much of the ultra-high energy influx is generated in ordinary galaxies and what proportion is due to particles emitted from radio galaxies and similar eruptive objects.

Clearly the detection of primary cosmic ray particles from galactic and extra-galactic objects is of importance – they will increase our knowledge of the kind of reactions and the previous and subsequent behaviour of the universe; but they are not the only particles that can assist the astronomer. In the last decade it has become realized to an increasing extent that neutrinos should also be able to provide much useful evidence, and the very difficult problem of observing them has been investigated.

Neutrinos, having no electric charge and zero mass – possessing nothing but spin – will pass through a star without hindrance unless, perhaps, they take part in some nuclear reaction. Most are thought to escape, since reactions involving them will only occur most infrequently. To the physicist they are one of the few stable nuclear particles, and because of this it is likely that they will travel great distances and through all kinds of material before they undergo a chance reaction. To give a simile, at least a hundred thousand billion billion (10^{23}) neutrinos pass through the human body during the normal span of life, yet of all these only one is likely to take part in an interaction with body tissue. Astronomically it has been pointed out that if the Sun emits neutrinos – and it is thought to eject a plentiful supply – then there should be no change of flux between midday and midnight, since it is the probability of occurrence and not the presence or absence of matter in the way that is the significant factor.

Neutrino reactions may be rare – indeed the neutrinos emitted from the large nuclear reactor at the Los Alamos Scientific Laboratory and known to be appearing at a rate thirty times greater than that expected from the Sun or other stars, were certainly observed, but the detecting equipment only recorded two reactions every hour. All the same the β-decay process (discussed in Chapter 5) produces many neutrinos. At the centre of the Sun where material is at present at a density equivalent to ten times that of lead and at a temperature of 15 million °K, an average proton will make 10^{16} collisions per second and, once out of every 10^{26} such collisions – in other words once every few thousand years – it will penetrate close enough to another proton with the possibility of forming a deuterium nucleus of two protons, with the emission of a positron and a neutrino. The deuterium will not form in every such case – rare though it is – and calculation gives a chance of one in ten million. However the number of protons is so great that the probability is that of all the energy generated within the core of the Sun, some ten per cent is irradiated in the form of neutrinos. Their emission is, therefore, far from uncertain, but it is so different a process from the emission of electromagnetic radiation that observing it would be tantamount to being able to 'see' right down to the central solar regions.

A theoretical analysis of solar neutrino emission indicates that two kinds of process will operate – the β-decay procedure (combination of two protons) and a reaction where two protons and a positron combine. This results in a 'continuous spectrum' emission due to the β-decay and also a distinct energy 'line' caused by the proton-proton-electron reaction. But the generation of neutrinos is not confined to stars of solar type; it is likely that bright hot stars are very powerful sources. Indeed, it seems that the hotter the source, the greater the emission of neutrinos compared with the emission of more familiar radiation. Calculation indicates that a population I red supergiant will emit ten thousand times more energy in the form of neutrinos, than in the form of electromagnetic radiation. In a nova the ratio of neutrinos to electromagnetic radiation is a million to one – an increase of 100 times: in a supernova a further increase of ten thousand times is expected, making neutrino emission ten billion times greater than the emission of electromagnetic radiation.

There is also the possibility that if neutrino emission is discovered to emanate from the most distant regions of space, this will be an indication that it was occurring at a very remote time; this could have significant cosmological repercussions, as we shall see in the final chapter. The problem, however, is to find some satisfactory method of observing these particles that possess neither mass nor charge, and which can only be detected on those rare occasions when they interact with other particles of matter.

In the Los Alamos experiment, using the heavy neutrino output from a nuclear reactor, Frederick Reines and Clyde Cowan employed a huge water tank, an inner tank of scintillation liquid and photomultiplier tubes, the neutrinos being detected when they reacted with the inner tank and caused scintillation. The danger here is that scintillation might also be caused by cosmic ray particles, but these would, at the same time, give rise to Čerenkov radiation at visible wavelengths, and additional electronics and photomultipliers deleted any such occurrences from the number of reactions being counted. Nevertheless neutrinos are expected products from cosmic ray cascades and Reines has now built a detector at a depth of 2000 feet in a salt mine near Cleveland, in the hope that all other particles will be shielded by the thickness of rock and earth above the equipment.

The scintillation liquid used is one in which there are neutrons present that are in a suitable state to capture neutrinos, and an isotope of chlorine – chlorine-37 – has been adopted. Its neutrons will react with a neutrino to form argon-37 and release an electron, provided the energy of the neutrino is a little more than 800,000 ev. Most solar neutrinos do not possess such an energy – at least those generated by the proton-proton reaction, whose energies lie around 500,000 ev – and if others are found that do, they will provide evidence for other more energetic nuclear reactions within the solar core. All the same this is important, and, if the high-energy neutrinos are observed, this will be the first direct observational evidence for processes that are only conjectured theoretically.

Utilizing chlorine-37 will still only be likely to produce a chance of encounter per chlorine nucleus of one to 4×10^{-35} per second, and since it is desirable to ensure a capture rate of the order of one per day, this means that the experimenter should use about 10^{30} chlorine atoms. If carbon tetrachloride (CCl_4) is used as the liquid, the bulk would be some 100,000 gallons or the equivalent of the load of forty road tankers: perhaps better, and certainly safer to use, is perchlorethylene (C_2Cl_4), where 100,000 gallons would contain some 2×10^{30} chlorine atoms for the same bulk. Plans for such a tank, immersed in a still larger tank of water, have been drawn up, and also a procedure for flushing the inner tank from any traces of argon. The argon-37 formed by any neutrinos is naturally radioactive, and will decay back into chlorine-37 in a period of some thirty-five days and thus, after a time, the number of argon-37 nuclei formed will reach a stable amount, the new nuclei replacing those that have decayed. The inner tank will, therefore, require cleansing at least every few months. Yet even with such an apparatus, it is still necessary to ensure that the neutrinos received are, in fact, from the Sun and not due to cosmic radiation or to any other cause. Very deep burial of the equipment is required, and from the experiments of Reines it would seem that 2000 feet is not low enough.

However, in spite of experimental difficulties and expense, it is clear that the detection of neutrinos from the Sun and their possible detection from other objects in space (although no technique for achieving this has yet been devised) would be of considerable assistance in determining the nuclear processes that are occurring in unobservable regions of stars and galaxies. Meanwhile, the increasing interest in the detection of primary cosmic rays seems likely to prove of considerable use in that it provides another new weapon in the astronomical armoury, and one that has its own particular contribution to make.

CHAPTER 8

Ultra-Short Wave Astronomy

Astronomy in the ultra-short wavelengths – the short ultra-violet, X-ray and γ-ray regions – cannot be carried out except from rockets or satellites, but the actual equipment used will depend to a great extent on whether the instruments are to be recovered or not. If they are to fall back to Earth, then recording on photographic film is practicable, but the equipment must be sturdily built if it is to survive the impact of landing, even though its fall is broken by a parachute. From rockets the observing time is limited – seconds or minutes are all that can be achieved – and for extended observations the artificial satellite is employed.

In ultra-violet observations there are a number of devices that are suitable, some for the intermediate range of wavelengths from 3400 Å to some 500 Å, and other for the extreme ultra-violet (usually abbreviated XUV) that goes as low as 40 Å. It will be convenient to start with the intermediate range equipment and then pass to those for the extreme end of the ultra-violet spectrum. In general few normal photographs are taken, since spectroscopic studies have received the most attention, but provided a reflecting telescope is used there are no problems over the filtering effects of glass nor the inability of a reflecting surface to operate successfully – indeed, the problem of reflectivity does not occur until one reaches the X-ray region. In the intermediate range, photographs of the Sun may be taken with quite a simple camera, and the spectroscope presents no inherent difficulties since reflection gratings (Chapter 2) are employed, except in the rare case when a grating is placed over the front of a rocket-borne telescope or a prism of calcium fluoride is used.

One kind of spectroscope much favoured by American investigators has been the grating spectrometer. Here the grating is curved, and after the radiation has travelled through a collecting lens of some substance such as quartz, or a collecting mirror, it is passed on to the grating and thence to the photographic plate. In one such design, used in the Aerobee-Hi rockets, two such spectrographs are mounted close together; the collector lens section deals with wavelengths from 3500 Å, while the other can disperse wavelengths as low as 500 Å. This is for a recoverable instrument, but where a satellite is being employed the instrument is modified into a grating spectrophotometer, since the observational results have to be radioed back, and this is most conveniently achieved by measuring intensities at various wavelengths. The intensity measuring devices are photomultipliers.

The observing instruments that may be packed into a satellite vary, depending on both weight and bulk. The first OAO (orbiting astronomical observatory) launched in April 1966 carried four eight inch diameter reflecting telescopes, one of sixteen inches diameter, two grating spectrophotometers plus appropriate filters, electronics, photomultipliers and various radiation counters. So far the largest practical design appears to be one in which a thirty-six inch diameter reflector is to be flown with appropriate spectrophotometric and ancillary equipment. The variations even for a small 'package' of this kind are legion: telescope-television cameras may be employed to map the heavens at different ultra-violet wavelengths, and there is a design using four twelve inch diameter telescopes with appropriate filters for this purpose. For the XUV range, the problems become a little more specialized, primarily because the experimenters wish to avoid the intermediate ultra-violet wavelengths on the one hand, and interference from γ- and X-radiation on the other. Again cameras and photomultipliers may be employed, but the photographic film requirements tend to be more stringent than for the intermediate wavelengths. The main photographic problem is that in the ordinary type of film – in which the photographic emulsion is suspended in a gelatin layer supported on a flexible transparent base (cellulose acetate or a plastic such as 'Mylar') – too much ultra-violet is absorbed by the gelatin layer at most XUV wavelengths longer than 40 Å. This may be overcome to a great extent by the use of Schumann-type emulsions, in which the gelatin is kept to a minimum and the surface layer of the films has a high concentration of light-sensitive chemicals, although at the extreme end of the XUV range, the ordinary X-ray film seems to be more suitable – here the gelatin content is higher but its power of absorption has passed its peak at wavelengths shorter than about 40 Å.

Cameras for the XUV range use no lenses but employ a pinhole to form the image. The pinhole, with a diameter of around six thousandths of an inch, admits very little light and, for the Sun, exposures of the order of three minutes are required – the rockets must therefore be stabilized for a period at least as long as this. In everyday use, the diffraction of visible light caused by a pinhole of this small size would limit the resolution of fine detail to 10′, but at wavelengths shorter than 500 Å, the diffraction effect is less noticeable and theoretically a one minute of arc resolution should be possible. Stabilizing and camera design problems limit the resolution in practice to about three times this amount. To filter away longer wavelengths and also to keep the photographic results free from irradiation by shorter wavelengths, two kinds of filtering are employed. The longer wavelengths are removed by a thin piece of aluminium foil about one ten thousandth of an inch thick, placed over the pinhole, but for the shorter a grazing incidence filter is used. This consists of two pieces of optically flat glass with their surfaces parallel and close together. Incoming radiation strikes one piece at an angle, is reflected to the other, and is then reflected into the pinhole. The radiation meets the glass at a grazing angle of about 10° and this is critical for the wavelengths employed – 170 Å to 700 Å, say. Shorter wavelengths would require a different grazing angle and are thereby filtered away. A pinhole camera may also be employed with a grating inside the camera body that reflects the radiation from the pinhole on to the photographic plate. Such cameras have been successfully flown in the Skylark rocket at the Woomera rocket range in Australia, and elsewhere.

For satellites, where photographic techniques are inconvenient, the spectrophotometer can be used if specially designed for the shorter wavelengths from 500 Å up and, in particular, the use of multiple gratings within the instrument can remove stray and unwanted emissions. The telemetering monochromator has also been employed in the United States experiments, and essentially this is an instrument which isolates a particular wavelength for study, usually by using a grating and two slits, and then passing the output to a detector of the photomultiplier type. Gas-filled counters like the Geiger-Müller, and ionization chambers may also be employed for the XUV and X-ray regions of the spectrum: admittedly these can do no more than count the number of photons being received from a given direction, and without the fine resolution possible with a stablized rocket and more directional equipment with a smaller acceptance angle for radiation, but since the stabilization devices weigh five times as much as a telemetering monochromator which, as its name implies, requires ancillary electronic equipment to telemeter its results back to Earth, the advantages are obvious. For preliminary surveys to provide first results in investigations and thus to supply information for the design requirements of more elaborate equipment, the simple counter is admirable.

The X-ray spectrum may be investigated by counters, but much work has recently been carried out on the design and construction of X-ray telescopes, particularly at Leicester University and University College, London. The X-ray telescope has a metal filter and operates by grazing incidence, although here the grazing angle is small, being no more than

4° for a wavelength* of 44 Å, 1½° for 8 Å and only 1° for 5 Å. The incident grazing radiation falls on a hollow tapered cylinder that acts in a way analogous to the main mirror of a reflecting telescope. Grazing incidence occurs on the inner wall of the cylinder, and the radiation then passes through an aperture at the focal plane of the cylindrical reflector (figure 15) to a detector. The one considerable source of error is that caused by the presence of a background 'radiation' of charged particles from space, since these will be counted by the detector. However they will possess the ability to cause scintillation in a suitable material, and the X-ray telescopes are surrounded by a plastic scintillating shield that

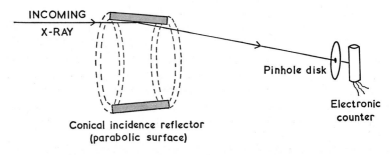

Figure 15

The principle of one form of X-ray telescope, whereby the incoming X-rays (of which only one is shown) are refracted by grazing the inside of a conical reflector. The rays then pass to an electronic counter by way of a pinhole in a metal disk which prevents stray radiation from reaching the counter.

is monitored by a separate detector, so that if both detectors – the one on the telescope and the other on the shield – register, that particular impulse can be electronically ignored.

To make investigations of the position of X-ray sources, the telescope must be as directional as possible. The Leicester University – London University telescope just described has an acceptance angle that varies with the aperture of the hole through which the reflected beam passes, and gives a field of view that may be as small as 2′. A further development is the use by Robert Boyd of a telescope with a specially curved second grazing reflector as well as the tapered cylindrical reflector. Analogous to the Cassegrain design of reflector used by the visual

---

* As mentioned in the second chapter, the classification of where the ultra-violet (XUV) ends and the X-ray region begins is somewhat arbitrary. The significant point is wavelength observed, since all the wavelengths concerned are due to high energy photons, whether they are called XUV or X-radiation.

astronomer, this should prove most valuable in differentiating one small X-ray source from another with better resolution than previously available.

A narrow acceptance beam has also been achieved by Herbert Gursky and a group in Massachusetts who have found it possible to measure an angular spread of 1', and this they have obtained by an ingenious system of two screens of parallel wires, devised by Professor Oda of the Massachusetts Institute of Technology. The wires – one hundredth of an inch thick and separated by a similar space – lie directly in front of a detector (figure 16), and when a parallel beam of radiation from a distant

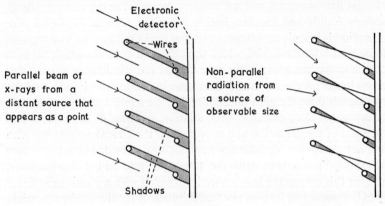

Figure 16

X-ray telescope utilizing two screens of parallel wires to detect a narrow beam of radiation. As a distant 'point' source of X-rays passes in front of the telescope the electronic detector receives a regularly fluctuating signal from shadows and spaces lit by radiation (left). If the source covers an observable area the ratio and sharpness of shadow to spaces lit by radiation is different and the electronic detector receives a different kind of signal.

starlike source passes across the wires, the detector is alternately irradiated and then in shadow: this causes the detector to emit a fluctuating electric signal. Should the source being observed cover an area of the sky, then the rays from it will not be parallel and the shadows cast by the front grid of wires will not be sharp, and the fluctuating signal will be different from that from the starlike source since it will not drop to zero and, for a particular size of source, the fluctuation will cease. This critical size depends not only on the diameter of the wires but also on the distance that separates the two screens. In the instrument constructed by Gursky, the critical size was 30'. In earlier work conducted by

Herbert Friedman and his colleagues at the Naval Research Laboratory in Washington, an acceptance angle of 10° was obtained by using a honeycomb of small metal tubes – rather similar to an automobile radiator – and with this device a considerable amount of very useful work was achieved. However, the demand now with X-ray telescopes, as with radio telescopes, is for increasingly fine resolution.

The incidence of cosmic particles can equally vitiate results in the γ-ray region and, once more, a scintillation check seems to provide the answer. Here the telescope can itself be a specialized scintillation counter of different design from the X-ray instrument. One proposal, made by William Kraushaar and George Clark, is to use a sandwich of two types of scintillation crystal and a Čerenkov detector for the telescope itself, sodium iodide and caesium iodide being the crystals employed. The operation is ingenious, for when photons strike the sandwich of crystals, these emit a scintillination which then dies out at a rate which depends on the substance used: in the case quoted the scintillation persists five times longer in the caesium iodide than in the sodium iodide, and the two signals from the photomultiplier that the scintillations produce may be separated electronically. The radiation, or some of it, passes round the sides of the sandwich and moves on to the Čerenkov detector with its photomultiplier. Cosmic ray particles – primary or secondary – may be expected not only to strike the 'telescope' head on but also sideways, and for this reason the instrument is provided with a plastic scintillator shield, observed by its own photomultipliers. If the shield, the sandwich, and the Čerenkov detector emit, then the signals are ignored, but if only the sandwich and the Čerenkov scintillator operate, the three pulses are transmitted. The strength of the two sandwich pulses, in addition to the Čerenkov pulse, allows the terrestrial observer to work out the direction of the incoming γ-radiation. Should γ-radiation cause the shield to scintillate, then its arrival will be ignored – but as this will only rarely happen, it is not likely to cause much distortion to the results. The acceptance angle of the telescope sandwich is some four times that of the Čerenkov detector and covers four per cent of the entire sky. Such an instrument may be expected to meet with a permanent background radiation that will also add now and again to the counting rate transmitted back to Earth, but such spurious results may be simply removed by allowing the instrument a period in which to be operated by this radiation alone, and subtracting the measurements obtained from the final results. Accordingly, when it is first launched, the front of the telescope is protected by a radiation-opaque screen of tungsten: once the background measurements have been made, the shield is then blown off on a signal from Earth control and the equipment is then fully operational.

Having sketched the types of equipment designed for investigating the ultra short wave ranges of the electromagnetic spectrum, some of the preliminary results can now be described, but it must not be forgotten that such results are preliminary. Observations of this kind are in their infancy, for the subject is almost a decade younger than radio astronomy – although the earliest results were obtained in 1946 – and much more may be expected as techniques improve and interpretations of the results become more securely based on previous experience.

It will be as well to begin with investigations of the planets, for although they do not themselves radiate at very short wavelengths, the atoms in the upper reaches of planetary atmospheres are affected by solar ultra short wave radiation; they are caused to emit such radiation by resonance with the incoming high energy ultra short wave photons. Moreover, they also scatter some of the ultra-violet radiation that reaches them, and they are frequently ionized and so emit due to this cause, while some are made to fluoresce.* Mercury gives no radiation of this kind – or at least none has so far been observed – but this is not surprising since the planet possesses no atmosphere. Venus, on the other hand, does provide some ultra short wave radiation, especially from atoms of hydrogen, oxygen, nitrogen, argon and carbon. The hydrogen emission is at a wavelength of 1215·67 Å – the Lyman-α line – that of nitrogen at 1200 Å and argon at 1048 Å. Oxygen atoms and some nitrogen atoms are found singly ionized, but the carbon is always molecular, and radiation has been observed from most of the kinds of molecules found in the terrestrial atmosphere.

By using artificial satellites, an examination may be made of the re-radiation of ultra-violet from the Earth and of other effects which are caused by solar radiation at these wavelengths. A 'day glow' has been discovered on the atmosphere over the sunlit hemisphere – this has a peak emission at 2130 Å and is due to re-radiation by nitrous oxide molecules: there are also other significant emissions due to ionized nitrogen atoms and to neutral nitrogen as well. Much incoming ultra-violet is, however, absorbed by oxygen that forms into a layer of ozone at a height of between nine and thirty miles above the ground. Such information as this is of considerable use meteorologically as well as from a purely geophysical point of view.

The body most closely studied in the ultra-short wavelength ranges is the Sun. Spectroscopic investigations have provided a great wealth

* Emitting a radiation of a longer wavelength than that received.

of information and, coupled with simultaneous observations in visual wavelengths, have resulted in a deeper understanding of the heat and energy distribution in the gaseous layers above the photosphere. The evidence on which this is based, is that XUV and X-ray photographs of the Sun taken from Aerobee and Skylark rockets show, as the wavelength is reduced, an increasingly larger solar disk compared with the visual image.

In the middle range ultra-violet, the longer wavelengths display a continuous spectrum crossed by dark lines, a spectrum that appears just like that of the photosphere with cooler gases lying above. However, when a wavelength of 2085 Å is reached, the picture begins to change. The dark lines become progressively weak, and, although the continuum persists, at 2000 Å, there is also some evidence of faint emission lines. By 1700 Å the emission lines are really bright, and it is clear that many substances like calcium are now ionized, and it is these that may be providing the continuum background. Observations made with the Skylark rocket between 2950 Å and down into the XUV range at 950 Å have allowed 300 emission lines to be mapped, and although not all are so far identified, three-quarters of the total have been shown to be caused by highly ionized substances, among them doubly ionized carbon, nitrogen with three of its seven electrons removed, and oxygen with four of its total of eight gone, while iron atoms are present with ten or eleven out of a total of twenty-six electrons ejected. Trebly ionized carbon is also to be found, and silicon with a loss of eleven of its fourteen electrons, both in the chromosphere which must, on this evidence alone, be very much hotter than the photosphere below it. A study of these observations also shows that although bright lines gradually replace the dark absorption lines of the everyday visual spectrum, the total amount of radiation available becomes increasingly less as the observed wavelength is reduced: half the Sun's brightness is concentrated in radiation between 7000 Å and 4000 Å – the visual range – but in the ultra-violet down to 1000 Å only another eleven per cent is to be found, while around 1000 Å only one millionth of the total output is emitted. But these figures are for the quiet Sun, and on occasions of enhanced activity the propositions alter considerably.

At the time of writing, the Orbiting Solar Observatories, OSO 3 and OSO 4, are still in action, and while their full results therefore cannot be quoted, it is of interest to note that these satellites can be commanded to transmit results at will. Astronomers at Harvard took advantage of this on the occasion of the prediction of a large solar flare – a prediction that arose from an analysis of visual observations. The results showed many photons from emission at the Lyman-α wavelength, and the fact that magnesium had lost nine, and silicon eleven, electrons. The Lyman-α radiation came from above the very active areas and the general evidence was of a considerably enhanced temperature of some 10,000°K, that is almost seventy per cent higher than the photosphere. The presence of trebly ionized oxygen was also observed higher still from a layer between the chromosphere and the corona, indicating a temperature of some 300,000°K. Up in the corona itself, observations in the XUV range at 625 Å provided evidence of a temperature of more than a million degrees Kelvin above the active areas, although, as mentioned in Chapter 3, the gas here is extremely rarefied and one should beware of thinking of temperature in straightforward everyday terms.

XUV observations from the Skylark rockets have confirmed this evidence, and made it clear that the energy generated in an active area is sufficient to ionize helium, while photographs taken in these wavelengths of the Sun's disk show details of the positions of the active regions with a resolution of not less than 2′, and probably as good as 1′. The latter are, of course, important for correlating with the photographs in visual light from Earth-based observatories. X-ray observations of the Sun have been made, and these confirm the observations in the XUV range, showing that there is a little emission under quiescent conditions but that this is enhanced over local areas, and photographs taken by Friedman's team as early as 1960 showed this very well, the corona clearly being the source of the major proportion of the radiation.

If the Sun is a source of ultra short wave radiation, then one may expect that other stars also emit in a similar way, and this seems to be generally so, although the extraordinary distances involved have prevented much information being obtained at present from normal stars. It has, however, been established that, as far as ultra-violet radiation is concerned, the strongest main sequence radiators are the very hot O and B type stars, which display high degrees of ionization. Moreover, observations of gaseous nebulae at visual wavelengths, especially the nebula in Orion in which O and B stars are embedded, confirm this, since the gas is stimulated to emit so brightly on visual wavelengths from the short wave energy received from the central stars. O and B stars may radiate a significant proportion of their output in the ultra short wavelengths, but there appear to be some stars that radiate virtually all their energy in the X-ray region: their nature presents a considerable problem to the astronomer.

These so-called 'X-ray stars' have been found in many constellations, but those in Centaurus, Cygnus and Scorpius have been more closely investigated than the remaining thirty or so known objects similar in kind. The Centaurus source has been studied by a team from Leicester University with rockets launched from Woomera, by the Australians and by an American team using rockets launched from Hawaii, and has shown considerable variation in emission. As Herbert Friedman has pointed out, of the thirty X-ray stars observed by him and his colleagues, many display variation, and they alter by as much as seventy-five per cent in less than one year: this may, of course, be a characteristic of all such sources which perhaps remain quiescent for a time and then suddenly burst into activity, but the Centaurus case is one that is worth describing a little more fully. On 4 April 1967 a team from Adelaide and Tasmania launched a Skylark rocket from Woomera which contained two identical X-ray systems and compared the Centaurus source with that in Scorpius. They found the Centaurus source to be a strong emitter of X-radiation, but at a longer wavelength than that in Scorpius. On 10 April 1967 the Leicester equipment was launched, also from Woomera, and again found the source to be intense; but five weeks later, when a rocket was despatched from Hawaii, the X-ray intensity had fallen to a value some six times less than that found from the Woomera measurements. A further check was clearly required, and in October 1967 another American rocket launching was made to examine the source: its intensity was found to be sixty times less than the Leicester measurements. It seems, then, that the Leicester observations were fortuitously made at a time of maximum X-radiation.

The Centaurus source shows no absorption of its X-radiation by interstellar dust or gas, yet this would be expected if the star were very distant. Moreover, it lies no more than $2\frac{1}{2}°$ from the central plane of the Galaxy, but if it were very distant, it would be likely to be masked by intervening material from this densely packed region, and, once again, this is taken to indicate that the object is relatively near. Kenneth Pounds and Arthur Meadows at Leicester believe that this evidence, coupled with the visual appearance of a very blue star in the area where the source is to be found, indicates a distance not smaller than 500 light years, and not larger than 2000.

The Scorpius source – usually known as Sco X-1 – appears as a blue star of apparent magnitude thirteen and, when observed with Oda's shadow technique telescope in the 1 Å to 10 Å range, launched on 8 March 1966 in an Aerobee-Hi rocket to a height of 100 miles, it was emitting X-rays far more energetically than any other ranges of the electromagnetic spectrum, so much so, in fact, that if the X-radiation were visible the star would appear to be of fifth magnitude – in other words some 1500 times brighter. Observations with the 200-inch tele-scope at Mount Palomar have revealed a noticeable variation at visual wavelengths that may amount to one magnitude (two and a half times) in twenty-four hours, and that even shows detectable changes every few minutes. However, these variations give evidence of being irregular. Changes have also been detected in the emission lines of its spectrum and in the source's colour.

The Cygnus X-ray star – Cyg X-2 – has also been found to be identi-fiable with a dim blue star displaying unusual features. Observations in visual wavelengths show not only a star-like image, but also a consider-able Doppler shift of the spectral lines. Measurements made by Margaret Burbidge and her colleagues at the Kitt Peak Observatory, Tucson, Arizona, indicated a movement of 160 miles per second, with a difference from one night to the next. Others, made later by Alan Sandage with the 200-inch gave Doppler shifts of some 600 miles per second.

With these facts to hand there has, naturally, been some disagreement about their interpretation, and it is clear that no one can yet be certain of the precise nature of the stars – for there is no doubt that in these cases it is some kind of star that is involved and not a distant galaxy or a gaseous nebula. One obvious possibility is the neutron star, mentioned in Chapter 5, but opinion seems to be veering away from this explana-tion, mainly because of the noticeable fluctuations in the emission of radiation. Another suggestion is that the objects observed are novae (not, it must be emphasized, supernovae – the visual effects would be far greater, especially with a supernova as close as X-ray stars seem to be): the sudden enhancement of radiation, particularly of X-rays, might conceivably be caused by the ejection of a hot shell of gas at supersonic speed. The temperature of the gas would have to be at ten million degrees to provide the thermal emission observed, for instance, in the Centaurus source, but this is quite likely in such an explosive situation. In addition, in so hot a gaseous envelope, the emission would for the most part lie in the X-ray region, while visual wavelengths would contain almost nothing but the radiation component from the original star. The team at Leicester favour this explanation for the Centaurus source, which they have been examining.

Recently a new explanation of X-ray stars has been proposed, mainly by Friedman, Margaret Burbidge and Sandage: they suggest that what are being observed are very close double stars. Burbidge and Sandage conjecture that if the two components have masses like that of the Sun, then the distance apart of their surfaces might be as small as one mile with a period of mutual revolution about each another of some ten hours. Such a close pair of stars would be very distorted, since they are so close, and they would be likely to trap interstellar dust and gas, that would then become heated. What is more, the speed of mutual rotation and the physical distortion could readily tear a layer of matter away from one, or perhaps both, stars – and this could occur more than once. It would be like peeling off part of the solar photosphere and disclosing the hotter gases underneath – X-radiation might then be expected to arise until the new surface cooled down to the level of the previous photosphere. Burbidge and Sandage not only make this tentative proposal for Cygnus X-2, but also for other X-ray stars, and they point out that as most lie close to the Galactic plane where young stars are to be expected (in the spiral arms), their distances appear to be close enough to make this likely (since if distant they could be observed only above or below the Galactic plane and the plane of the arms), while it has been suggested from time to time that stars are mostly (if not entirely) formed in pairs that may, or may not, separate later.

Friedman, on the other hand, has suggested that although a binary system is involved, one of the stars is a white dwarf, with a consequent powerful gravitational attraction for interstellar matter. Such a mass of gas, if ionized, would form a swiftly moving plasma that could well reach a high temperature – perhaps of the order of fifty million degrees – and copious X-ray production would result. Both this suggestion and that of Burbidge would lead to the emission of a quickly changing quantity of radiation, but it is too early yet to say whether all X-ray stars are of the same type. Novae may be involved; in the case of Sco X-1, photographs made in 1896 have been found to include the source, looking then merely like an unexceptional blue star, and this makes the nova explanation appear more likely in this instance, although probably not for Cyg X-2.

The sources of ultra short wave emission so far considered are those in which the radiation is emitted by thermal action of electron jumps within the atom. There are, as discussed in Chapter 5, a number of non-thermal possibilities of energy generation in these wavelengths, and it seems likely that these may be the cause of some of the radiation received from interstellar gas, where this is associated with a very strong magnetic field. Here synchrotron radiation and collisions of high speed nuclear particles with gas atoms may emit ultra-violet and even X- and γ-radiation, but at present it is discrete sources that have been observed, rather than much of the inter-stellar medium, and where the latter has been observed the ultra short wavelength radiation seems to have been generated by hot stars close to or embedded in it. One interesting fact has, however, come from studies of the absorbing power of interstellar dust at these short wavelengths. This is because the grains, which are conglomerates of molecules – generally thought to be ice with some metallic impurities – will have an absorbing power that varies at different wavelengths according to the chemical composition and grain size. At optical wavelengths, some information can be extracted, but at shorter wavelengths this would appear not to be so since in the ultra-violet range the absorbing power is increased. Nevertheless, the ultra-violet has offered a significant clue for, at wavelengths below 2500 Å there appears a fluctuation in absorbing power, and this has led Nalin Wickramsinghe to suggest that many grains may be elongated and composed of flakes of graphite.

In the X-ray wavelengths one powerful and discrete source within the Galaxy is the Crab nebula. This supernova remnant radiates strongly in radio and visual wavelengths, as mentioned earlier, and its X-radiation has been closely studied, advantage being taken of an occultation of the nebula by the Moon in July 1964. An Aerobee rocket was launched from the White Sands missile range at the time of the occultation and counts of X-radiation taken during the gradual progress of the Moon's limb across the nebula. The rate of change of the X-ray photons received reached a maximum nearly four minutes after launching and then slowed down, and this has been taken to show that the angular size of the X-ray source within the nebula is no more than 1', or about one third of the size of the visual image. The source is too large for it to be caused by a neutron star – or any other star for that matter – and the strong emission must therefore be due to some other cause. The synchrotron process has been considered, but the intensity of the X-ray emission received would mean that electrons with an energy of 30 Gev would be required to move through the central regions of the nebula, and while this is not impossible, the fact remains that such electrons would lose energy quickly. After some thirty years, the energy would drop to half the original value, but the supernova explosion is known to have occurred more than 910 years ago, so unless some continual influx of very high speed electrons can be discovered, this explanation is not acceptable. Recent observations by George Clark of the Massachusetts Institute of Technology have shown that the Crab emits not only high energy X-radiation, but also X-radiation with an energy of only 10-50 Kev, and the life of electrons with this energy is far less than for those in the Gev range: its average is one year for a fall to half the original energy. Clark's observation makes the problem of X-radiation even more acute, and thermal X-radiation issuing from a central star to provide the observed intensity does not seem likely, since the temperature of the star's surface gases would need to be several hundred million degrees even for the Kev energies. Possibly the central (exploded) star may have some mechanism for injecting high energy electrons into the gas surrounding it, but what kind of mechanism this could be has so far defied explanation.

However, there is one suggestion that holds promise, and this is connected with the probability that, in a large supernova explosion, the nuclei of heavier atoms are created. Many such atoms have unstable isotopes, and many such isotopes may be generated with, as is already known, lives before decay that would certainly embrace the time since the original explosion occurred. It is at present impossible to be certain about this explanation, and a number of astronomers including Margaret and Geoffrey Burbidge, Donald Clayton, William Fowler and Fred Hoyle have been investigating the problem to discover, in theory, what elements are likely to be produced and what decay reactions might be expected to generate in the way of ultra short wave radiation. Preliminary results indicate that there should be a γ-ray flux as well as one of X-rays, and there is clearly a need for a γ-ray investigation of the Crab to be undertaken, since this would provide not only evidence for the acceptance or rejection of this hypothesis, but also act as a guide for the formulation of other explanations.

Some X-ray sources have been discovered lying far away from the plane of the Galaxy, and it may well be that these are extra-Galactic sources. Indeed, it is clear that two are certainly within $0 \cdot 1°$ of the direction of two well-known and peculiar galaxies – the radio galaxy Cygnus A, and the radio galaxy Virgo A, known visually as M 87. The latter is of particular interest, as although its X-radiation is some 200 times less intense than, for instance, that from Sco X-1, the latter is not more than 2000 light years distant (and probably closer than this) while the distance of M 87 is thirty-three million light years. The reader may recollect that not only is M 87 a strong emitter at radio wavelengths, but that it also possesses a blue-coloured jet of material in two sections (plate VIc). The light from the jet is polarized and the jet itself is some 3500 light years long and about 300 light years thick: its composition is uncertain but may possibly consist of clouds of fast protons and electrons: certainly the polarization suggests synchrotron radiation at visual and other wavelengths. If this is so, then very high energy electrons could be produced by proton collisions in the jet, and could be provided with energies sufficient to give synchrotron X-rays. The main source of the X-radiation would, however, appear to come from the very dense core of stars –

denser by half a million times than stellar space in the neighbourhood of the Sun – where there is considerable energy in the photons of visible wavelengths which could create X-radiation on collision with high speed electrons, perhaps from the inner parts of the jet.

A third extra-galactic X-ray source is the quasar 3C 273 (plate XII): the visual similarity of this object and M 87 is self-evident. The radio components are centred on the jet and the stellar-like core, but the latter may be undergoing a series of explosions, since there seem to be four radio sources which, together, make up the second radio component. Spectroscopic observations show many changes within the main body and these occur quickly, taking no more than a few weeks. Herbert Friedman has studied this object in detail, has noticed that it appears to be becoming gradually dimmer, and computes that this would mean that it will reach the brightness of a normal galaxy in something like 3000 years. He has also made the suggestion that the 'main body' has, in fact, been thrown from the jet as the result of an explosion – hence its four radio emission centres – but not the centre of the present quasar activity. The X-radiation from 3C 273 is some fifty times greater than the radio luminosity and it is also a source of intense, but decreasing, infra-red radiation that is still greater than that at any other wavelengths. Shklovsky has suggested that the visual radiation arises from collisions between photons of infra-red and very short wave (millimetre) radio radiation, and primary cosmic ray particles, while Friedman has pointed out that such a mechanism would generate X-radiation of the kind observed, and he has predicted that X-radiation is to be expected from Seyfert galaxies. Techniques using improved X-ray telescopes within orbiting satellites or at a lunar observatory will doubtless prove, or disprove, this suggestion in due course.

Lastly, it must be mentioned that there is evidence of a weak 'background' emission of X-rays in the universe, evidence that has been confirmed by recent rocket-borne X-ray detector flights. Friedman suggests that if, as seems likely, primary rays pervade all space, and if the density of intergalactic gas is only one gramme per one and a half million cubic light years – a figure that does not seem unreasonable – then the background X-radiation could be produced by the collision of the cosmic ray primaries with the radio radiation photons that are present due to the background of radio radiation that has been observed. This seems to beg the question of the generation of the background radio radiation – for which the background X-radiation may be taken as additional evidence – but the explanation of its presence depends on cosmological implications that will become clear in the subsequent chapters.

CHAPTER 9

The New Universe

The universe as it appeared to the visual astronomer before the advent of observations in the invisible ranges of the spectrum, and of radiation generated by non-thermal processes, has already been sketched. It presented a somewhat static picture, such changes as appeared to exist being for the most part slow and gradual – only the novae and supernovae demonstrating anything in the nature of a catastrophe. The galaxies were found to be moving apart, and astronomers had reached some tentative conclusions about their possible evolution: stars within the galaxies were believed to develop, and the original direction of evolution from hot to cold was reversed after the proposal that thermonuclear reactions are the source of stellar energy. As a whole, there was the concept of an expanding universe, using modifications of Newton's theory of gravitation and the theory of general relativity to produce an idea of the past and future state of the universe. The new observations cannot do less than cause a revision of some of the details, but there is an indication that a more radical re-thinking of the whole picture is necessary.

First of all, observations in the radio, XUV and X-ray regions make it abundantly clear that in some places, both inside and outside our Galaxy, catastrophic changes occur. The discovery of X-ray stars underlines the catastrophic element, for either they are novae or distorted erupting binaries – or both – unless perhaps they are the result of a catastrophic explosion of an old star to form a neutron or hyperon star. With the radio galaxies and the discovery of Seyfert galaxies and quasars, the simple previous classification of galaxies into spirals, ellipticals and irregulars needs to be extended, and the possibility considered that the types or species of this classification form less of a sequence than has hitherto been supposed.

In the second place, observations in the new wavelengths and the discovery of the mechanism of synchrotron radiation make it clear that interstellar and intergalactic material, together with their magnetic fields, play a part that is more significant than that indicated by observations at visual wavelengths. In brief, the modern universe appears to be com-

posed less of separate species of bodies, each with its own specific func-
tion, than a dynamic concourse of bodies in various states of development or decay, and different only in the quantity of nuclear particles
composing them. It seems, in fact, as if the universe is something that is
undergoing continual change, and this has important implications on
our general outlook of the universe as a whole.

On the largest scale it is necessary to find some overall picture that
will account for the different extra-galactic objects, and find not only an
evolutionary pattern embracing elliptical, spiral and irregular galaxies,
but one that can place radio and X-ray observations of galactic eruptions
into the same scheme. To achieve anything along these lines means that
a decision must be made about the red-shift of quasars – are these truly
a measure of velocity and, if they are, may they be interpreted on the
basis of Hubble's hypothesis that the most distant objects are those
with the greatest velocities? Since 1933 a number of attempts have been
made to explain the red-shifts as due primarily to some factor other than
a recession in the line of sight. E. A. Milne, in a theoretical universe that
he constructed from a few basic assumptions, was led to suggest that the
red-shift partly arose from the difference in rate between the basic
behaviour of atomic interactions and that which we observe in our own
local part of the universe – the solar system. He postulated, in fact, two
time scales, both running at different rates, the red-shift being a consequence of the apparent slowness of distant atomic behaviour where we
are observing back in time when the time rates were very different, compared with that behaviour as we measure it now in the laboratory (for
it is in the laboratory that the astrophysicist determines what his standard
positions of spectral lines shall be). Milne's 'kinematical relativity' has
not been able to withstand the many criticisms levelled against it, not
least the fact that it rejects the general theory of relativity which contains
a satisfactory explanation of gravitation, and is unable to fill the vacuum
this has caused. Finlay-Freundlich suggested that the shift could be
explained by a scattering of radiation by intergalactic dust – this would
cause images to be reddened in a way analogous to the reddening of the
Sun by dust in the terrestrial atmosphere, and provide an apparent shift
of spectral lines. Unfortunately the scattering involved would affect the
images of the bodies observed, and would render it impossible to obtain
a sharp unblurred photograph: the explanation due to scattering is self-
evidently untrue.

These, and other proposals that have been made, have not found
favour in the astronomical world, but because there is no agreed alternative to the velocity explanation, this does not prove that the latter is

correct. However there are positive reasons for accepting the shift as a velocity of recession: on the one hand it is found to exist in local parts of space and the rotation of the Sun can, for instance, be inferred from other observations; and, on the other, it should be a proportional effect since it affects each spectral line by a percentage that depends upon its wavelength. Such a proportion is observed in extra-galactic red-shifts.

The universe certainly gives every evidence of expanding and the question naturally arises, if we move back in time surely a contraction should be observed? And, if we extrapolate sufficiently far, then it would appear that all matter was concentrated together in one place. Incidentally, since space, time, and matter are inextricably interwoven according to relativity theory, such a concentration of all matter would mean also a contraction of space and an analogous effect on time – in other words the universe would be a point and no more; what the mathematician terms a 'singularity'. This leads, of course, to the assumption that the universe began at some definite time in the past and expanded from a singular state – a condition often likened to a primeval explosion, so that the theory has frequently been referred to as the 'big bang'.

Calculation gives some tens of thousands of millions of years ago for the beginning of the expansion, and various computations have been made of the kind of state in which matter would be when concentrated into a 'super-atom'. George Gamow and his colleagues in the United States have suggested that, at the commencement, there was a concentrated mass of neutrons many of which disintegrated to form protons and electrons, and that in something of the order of half an hour the atoms of heavy elements were synthesized. However, a detailed analysis shows that it appears possible that, although hydrogen and helium could have been synthesized, heavier elements would have had to form at a later stage. Possibly they did so in the local condensations that we now observe as galaxies which, owing to the disruption of the primeval atom, were put into a motion of recession. However, the late Abbé Lemaître postulated a greater age and a primeval atom of considerable complexity, together with a disintegration that produced atoms no longer in existence but with immense nuclei that themselves decayed to produce the elements we now observe. In brief, Gamow's theory supposes primeval thermonuclear fusion, and Lemaître's nuclear disintegration.

The observations in the invisible ranges of the spectrum clearly indicate that considerable nuclear activity is taking place in the Sun and in other stars – in some bodies like X-ray stars in a very energetic and catastrophic way – and it would seem that this is where the synthesis of the heavier elements occurred, and is still occurring. Indeed, the whole concept of stellar evolution as derived from observation in every wavelength supports this view and, in consequence, if the big bang hypothesis has any validity – and it is probably true to say that the majority of astronomers believe it does – then the early stages are likely to be those described by Gamow.

There is, however, another interpretation of the universe, which was briefly mentioned in Chapter 1 when the perfect cosmological principle was discussed. This, the steady-state theory, propounded in 1948 by Herman Bondi, Thomas Gold and Fred Hoyle, at a time when it appeared that the date of the big bang explosion worked out (from observations of extra-galactic red-shift) to a value smaller than that computed for the age of our Galaxy, was originally designed to overcome this serious anomaly. In 1952 the anomaly was removed by the realization that the distance red-shift relationship was an underestimate: optical observations made this abundantly clear. Nevertheless the theory continued to be developed, and although it postulates that the universe as a whole is in a steady-state, always presenting the same general appearance to an observer wherever he may be and whenever he may be observing, it also allows change. Thus, if we now see galaxies being born, living and dying, then so will astronomers throughout the existence of the universe – the overall picture remains the same.

The continual recession of material due to expansion will, on the face of it, completely destroy the hypothesis, since space is continually being depleted, but the proponents of the theory have suggested that matter is continuously being created in the universe. Precisely how this is occurring is not specified, but the result is the appearance of hydrogen nuclei at a slow rate – about three hydrogen nuclei per cubic inch per annum. The creation of matter means the creation of energy, since relativity theory has shown the equivalence of matter and energy, and it appears to contravene the 'law' of the conservation of energy with which the physicist is familiar and to which he is wedded. The steady-state cosmologist denies this on the count that the continuously created material merely replaces that which disappears over the visible boundary of the universe, and claims – by implication – that such disappearing material ceases to be valid for consideration on the grounds that the unobservable is outside the purview of science. And, of course, there is always the possibility that the law of the conservation of energy may be proved invalid – at least on the very largest scale – although it is only proper to realize that no evidence has yet been adduced to lead physicists to seek such a revision. Indeed, the physicist has all along been suspicious of continuous creation, mainly because no explanation whatsoever has been suggested for the way in which this can happen, but if the originators of the steady-state theory made no proposals originally, this is not so now. Hoyle has recently suggested a creation field, while both Pascual Jordan and William McCrea have provided some rather more physical explanations, Jordan proposing that such material appears as supernovae and McCrea that it is generated within galaxies.

There is another physical law that must also be remembered in any consideration of the entire universe, and this is the second law of thermodynamics, concerned, as its name implies, with matters of heat and energy. Originally derived to explain phenomena of heat generation and dispersal in machinery, with particular reference to the steam engine, it contains the simple principle that a hotter body will transfer heat to a cooler one, until such time as both bodies are at the same temperature. At this stage, provided the bodies are isolated from any external influences, the exchange will cease since there is nothing to cause any further change: the original difference between them will be evened out.* The process is irreversible and clearly has a far wider field of application than mechanical engineering; indeed James Jeans and Arthur Eddington, who did so much for cosmology in the years before the Second World War, often referred to the 'heat death' of the universe, envisaging a time when all the radiating bodies in space would have spent their energy in bringing about a uniform temperature. The only way out would appear to be the influx of energy in a new form and, with the equivalence of matter and energy in relativity theory, the continuous creation of matter would provide just such an influx. The universe would be continually rejuvenated and need, therefore, possess no end: there is no reason either why its beginning (if it had one) could not be placed an infinite time in the past.

The steady-state theory demolishes the problem of the possible heat death of the universe if this problem really exists, for it must be emphasized that it has never been proved that the second law of thermodynamics is applicable to the universe. Only if the universe may be considered as an isolated system can the law be used and, even if this view is taken, it seems to many cosmologists rather too sweeping an assumption to consider that there may be no means of reversal of heat distribution. In the nineteenth century Clerk Maxwell convincingly

---

* As heat is transferred, the two bodies will become more uniform and the uniformity or 'entropy' will increase. It becomes a maximum when both are at the same temperature.

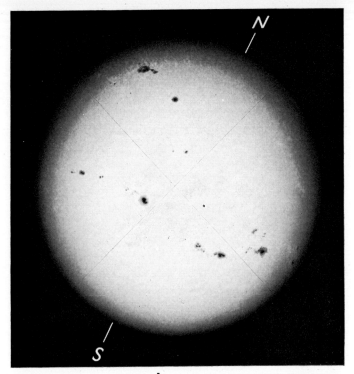

Xᴀ. The sun photographed ordinarily in daylight (i.e. integrated light of all visible wavelengths).

Xʙ. The sun photographed in hydrogen light (Hα), showing a flare on the north-west limb (north is at the top).

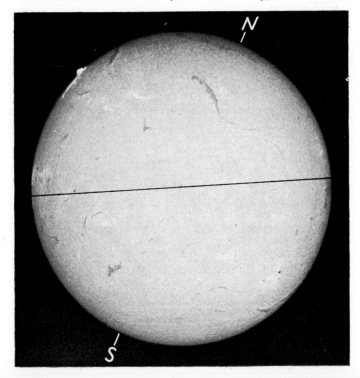

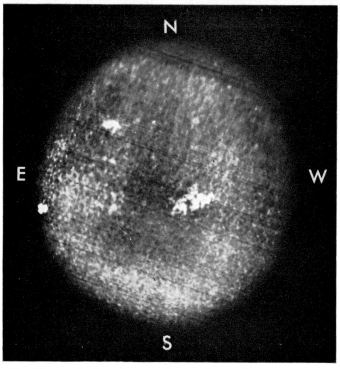

Xc. The sun in the calcium and potassium radiation, the bright patches showing these areas which were active at the time that the photograph was taken.

Xd. The X-ray sun taken from a rocket at the same time as photograph (c) was exposed. The bright patches are those areas from which X-rays are being emitted most intensely.

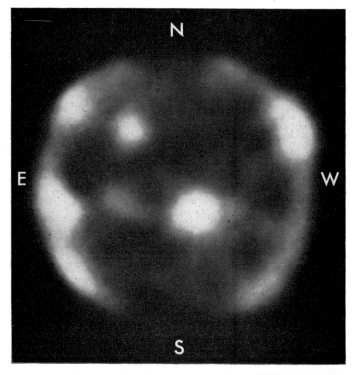

argued that a process of reversal could be envisaged, and although his argument was in some senses a fantasy (he referred to a demon re-ordering the statistical behaviour of molecules whose motion gave the effect of heat), it nevertheless established the principle that the second law is not infallible. Possibly the observations now being made in the extreme ranges of the spectrum are indicative of processes that may provide the kind of influx needed, although it is too early yet to do more than say that the question is an open one; but certainly cosmologists do not feel the second law to have a rigorous stranglehold on concepts of an infinite universe, as Jeans and Eddington did.

But if there are doubts about the validity of the second law of thermo-dynamics when applied to the entire universe, there appears little doubt about the universality of gravitation. This force, the formulation of which was first prepared by Isaac Newton, has a unique property among all physical forces: it cannot be nullified. The writer of science fiction has often availed himself of anti-gravity devices, but there is no physical reason to suppose that these could ever exist and, indeed, every evidence to presume that they cannot. In the nuclei of atoms other forces are more powerful over the short distances and with the small masses involved, but there is no suggestion that gravity ceases to operate there. Electric forces – an inherent part of many nuclear particles – affect only those particles that possess positive or negative electric charges, but gravitation affects all. It even affects those like the neutrino which have no mass, since they possess energy and, on relativity theory, energy interacts with gravitation: the deflection of photons of starlight by the Sun's gravitational field is an observed case of this kind. In the general theory of relativity, gravitation is interwoven with the very substance of the universe, with space and time, and in this it is a more significant factor than it was in Newton's cosmology. In formulating his laws of behaviour of planetary and terrestrial motion, Newton referred to the movement of a body as uniform motion in a straight line, unless the body is acted upon by an external force of some kind. But the universe is far from empty – observations in the invisible wavelengths are beginning to show how much previously unsuspected matter there is – and no body can, therefore, ever travel in a straight line; it must always travel in a curve. General relativity takes account of this fact, and removes the straight line from cosmological consideration by the simple expedient of using a curved space geometry. Space is distorted from the customary Euclidean space of simple everyday geometry by gravitation – by, in fact, the presence of matter. Since the discovery of quasars, and the need to find some mechanism that can cause the emission of such great quantities of energy, the question of the nature and effects of gravitation have been subject to some careful examination, and, in particular, the phenomenon of gravitational collapse.

Essentially, gravitational collapse is the tendency bodies have of falling inwards to the centre – to the centre where their gravitational attraction is concentrated. It is this that prevents the Sun from exploding, as it would do if it were not for the tendency of the mass of gas composing it to fall inwards and so balance the outward pressure caused by energy generation in its central core. Gravitational collapse is the cause of condensation of stars or of galaxies from a mass of gas, and it is the tendency of matter to congregate together that in theory can result in the catastrophic formation of a neutron star. The existence of much that is found in the universe is, indeed, explained by the balance between the force of gravitational infall towards the local centre of gravity, and the pressure outwards due to thermal and non-thermal* effects.

No star, it is thought, can hope to balance gravitational collapse by thermal radiation for ever, whether the second law of thermodynamics applies to the whole universe or not, for radiation is continually being lost by emission into space, as evinced by a star continuing to shine. A time is bound to come when, as already sketched in Chapter 4, a star begins to collapse. For a while, new thermonuclear processes may halt the contraction, but in the end it will continue until, if a star is not large, it ends up as a body in which non-thermal pressure balances the tendency to gravitational collapse. A white dwarf star results or, if the mass is much smaller, a cold body like the Earth. On the other hand a star may have such an internal distribution of matter that at some stage the thermonuclear processes take over and the star explodes, producing a supernova outburst. In addition there is the possibility discussed in Chapter 5, that a super condensation takes place and the body finishes its existence as an emitter of radiation by turning into a neutron star or an hyperon star, objects whose existence is not yet supported by any certain observational proof, but in which non-thermal pressure would still be sufficient to prevent complete gravitational collapse. According to Karl Schwarzschild, Robert Oppenheimer and Hartland Snyder, relativity theory shows that there is a minimum radius for any given mass at which gravitational collapse will overcome the non-thermal pressure, due to the concentration of matter in so small a volume. The radius, sometimes called the 'Schwarzschild radius', is small – for a mass

* Non-thermal pressure is that due to the nuclear forces within an atom or the force exerted by electrons in orbit around a nucleus.

as large as the Sun it would be no more than two miles – but for larger bodies the radius is correspondingly greater, and in general the opinion is that only large stars will have enough mass to compress matter sufficiently against the non-thermal pressure to permit gravitational collapse to occur.

Should gravitational collapse happen, then theory indicates that a strange series of consequences will follow. First of all, before the Schwarzschild radius is reached, the surface of the star will still be hot enough to emit photons, then, at what is termed the 'first critical stage' (when the radius of the body is some one and a half times the Schwarzschild radius) some photons will be pulled back by gravitation into curved paths around the body, and it will possess a photon shell composed of those photons emitted at a tangent to the stellar surface. Some of these will gradually leak away from the shell and continue to do so for what, theory indicates, will be an infinite time. As collapse continues to the Schwarzschild radius, more photons are emitted but more are captured. The 'second critical stage' occurs when the Schwarzschild radius is reached, and then only photons leaving the surface perpendicularly can escape, but they lose energy as they do so with the result that the cloud that they form, and from which some will leak, will be redder in colour than the first: it will also be smaller, but still just a little larger than the Schwarzschild limit. To a distant observer, then, a star undergoing gravitational collapse would appear to possess a hovering cloud and, at the same time, the star itself would dim until it disappears, leaving only the second cloud to add its reddened light to the first. The collapse is very fast, the speed depending on the mass of the body involved, being longer the greater the mass since, according to relativity, the heavier the body the more time is dilated and the longer a process appears to take.

The 'third critical stage' occurs when all the matter and unescaped photons, now in a radius smaller than the Schwarzschild limit, are pulled still further together. After the Schwarzschild limit was reached, the body would lose all contact with other observers in the universe because the density would be sufficiently great to cause space-time to become folded round on it. Yet computation shows that, from theoretical considerations, the collapse continues until the enclosed 'piece' of space-time become 'singular' – in other words reaches a point when it collapses in on itself. The mass, the photons, and the space around them have vanished and all that will remain is a distorted funnel-shaped depression in space (or more properly in 'space-time'). The final stages – including the collapse to singularity – are unobservable outside, distant astronomers seeing nothing but the critical stages one and two, already described.

It is the interwoven nature of space and time in general relativity that causes the perpetual observation of the penultimate stage of the collapsing star, but it is not this that worries the physicist about this whole hypothesis – he feels that something is seriously wrong when he is asked to consider the virtual crushing and vanishing of matter without the transformation of matter into energy with which he is now familiar. Perhaps, he conjectures, there is a fault in the reasoning, or in the theory. when one pushes laws of behaviour to their extreme – the results of doing so may well be invalid. Such a view has been taken by John Wheeler at Princeton, who has suggested that the problem may be overcome by supposing that once the Schwarzschild radius is reached, the gravitational pull behaves in discrete jumps with a limiting minimum. He has made his suggestion by drawing an analogy between two early theories of the atom – Rutherford's, where there appeared to be no reason why the orbiting electron should not fall back into the nucleus, and Bohr's theory, where every electron had a definite number of orbits in which it might travel and in which its collapse downwards emitted energy in discrete steps, although it could never drop down as low as the nucleus. Such a discrete or 'quantum' theory of gravity has not yet been completed, but it would appear likely to modify the space-time conditions near the singularity.

But Wheeler's solution is not the only possibility and a different approach has been taken by Hoyle and Jayant Narlikar, who point out that another of the unique features of gravitation is that when a system collapses under a gravitational pull, energy is given out and, in addition, the force of the collapse increases as the matter is crushed more closely together. In other words, the energy emitted during a contraction of a body like a star does not, as one might expect, lead to a diminution of the gravitational pull – as is found with other forms of energy – but to an increase. Hoyle and Narlikar conclude, therefore, that a new way of looking at gravitation is required, and that the most convenient is to regard it as a form of negative energy. As something positive (the emission of energy) is extracted from it, the negative energy is increased and the body contracts. The approach to singularity in gravitational collapse might therefore be arrested by some other form of negative energy that grows more rapidly than the gravitation does, and, by a repelling action, arrests the catastrophe. Hoyle and Narlikar believe that such a negative energy effect has been discovered, and that it is in a field of negative energy that the continuous creation of matter occurs. While the reader may question whether continuous creation has been discovered rather than merely postulated, the concept is of some significance, and worth pursuing further.

The idea of continuous creation, as mentioned previously, has often been considered unsatisfactory from a physical point of view and has led to the suggested alternatives of Jordan and McCrea. Hoyle and Narlikar have stated that it is impossible that the matter being continuously created can arise either from radiation or from any nuclear cause – any source of positive energy as they term it. Such sources would be likely to become exhausted very soon, and the expansion of the universe would appear to make it improbable that creation could continue at a constant rate. Hoyle has therefore suggested that matter is created in a negative energy field which is continually made less negative by expansion, but more negative by the generation of matter. In effect, the outcome of expansion is balanced by the generation of matter and a steady-state results. Although the creation of matter in this negative field (C-field or 'creation field') is not explained, the concept has a more immediate application to observed phenomena, since Hoyle has suggested that the immense energy and red-shifts observed in quasars are caused by the gravitational collapse of a large body.

An analysis of the red-shifts observed and their explanation in terms of gravitational collapse has been conducted by Kip Thorne and William Ames of the California Institute of Technology. They find that photons emitted at the first critical stage should show only a small change in wavelength, but those from the second critical stage should give a single broad line centred at about three times the emitted wavelength. These figures appear rather large, except perhaps for the quasar 1116+12 and, moreover, one would hardly suppose the displacement to be due entirely to a gravitational shift – a Doppler (recession) shift of some magnitude is to be expected. However the gravitational collapse hypothesis cannot be discarded solely on this account, for the theoretical considerations have been computed on the assumption that the collapsing object is perfectly spherical and that it is not rotating. Yet galaxies are rarely, if ever, perfectly spherical, nor are stars, and none are known in which rotation and internal movements are absent; but, from the theoretical physicist's point of view, the trouble is that if these points are taken into account, the problem of calculating the results becomes extremely difficult. An insight into the changed situation that such factors would bring about can be obtained from a theorem about space-time and the motion of photons proved in 1964 by Roger Penrose of Birkbeck College, London. Although the question is complex, it appears that either there could be a singularity or, owing to a 'hole' in space-time, the extra-ordinarily dense matter could be squeezed out to reappear elsewhere, either in our own universe, or in some mathematically possible universe with its own separate space-time that impinges on ours at what would otherwise be the singularity.

Fantastic though this kind of explanation appears, others have inde-pendently computed mathematical examples of collapse providing similar results, and the reader is reminded that much that appears strange is bound to occur in a universe in which the common-sense world of terrestrial experience has been transformed into the space-time cosmos of general relativity. Strictly, Penrose's theorem is limited to an infinite universe, but although there is no rigid proof yet for the con-ditions obtaining in a universe of finite extent – which seems to be the kind of universe which we observe* – there are indications that the results of collapse would not be very different. Yet one is left with an uncomfortable feeling that gravitational collapse has more to be said for it mathematically than physically. Perhaps here we have a phenomenon occurring in an intense gravitational field which requires a new physics for its explanation – a view that a study of the history of astronomy could lead us to expect and which Hoyle seems now to feel convinced is so from an analysis of astronomical and anti-physical evidence.

By now it will be clear that the question of the nature of quasars is still unsettled. Some show more than one red-shift, one quasar (P and S 0237+23) providing nine different values, and one can make no definite pronouncement about their distance. Most astronomers seem to favour the view that the distances are very large – thousands of millions of light years – but if they are examples of gravitational collapse, then this cannot be so: they will be likely to lie outside our Galaxy but not at prodigious distances – only some hundreds of millions of light years. Perhaps the desire to interpret the quasars as very remote objects is based, at least subconsciously, on a wish to promote the big bang theory of the origin of the universe rather than the steady-state. Theoretically there is little to choose between these possibilities – one can indeed claim some considerable elegance for the steady-state proposals – but the con-

* It is generally accepted that the universe is 'finite but unbounded', a concept that is impossible to draw and difficult to conceive since it involves four dimensions – three spatial and one temporal. Perhaps the best one can do is to take a two dimen-sional analogy using the surface of the Earth. Here there is a spherical surface on which the ordinary laws of Euclidean geometry do not apply – for instance, the in-ternal angles of a triangle do not add up to a maximum of 180°, but reach higher values – and we have a surface to which there is no boundary although its area is finite in extent and may be precisely calculated. The finite but unbounded universe may, likewise, be expressed with mathematical, if not pictorial, precision.

tinuous creation of matter disturbs the physicist, although he seems quite happy to accept the sudden appearance of a primeval atom at some remote epoch. In such a situation it is obviously necessary to turn to observation and see what clues this can provide, although we must not hope for a definite answer. At the moment it is impossible to be certain one way or the other, even though some of the indications will be discussed in the next chapter, but even if the steady-state, say, is rejected or seems likely to go, this does not leave the big bang in sole possession of the field. There is another important possibility – the oscillating universe.

Using the equations of general relativity, it is possible to prepare a 'model' of the universe in which the radius increases, as in the straightforward expanding models, but only does so as far as a limiting value; after this the universe begins to contract. The total time for such a cycle, from the initial expansion back to a concentrated state, is some 60,000 million years, and during expansion the distant galaxies produce a red-shift – as they are now observed to do – but during the contracting phase, a violet-shift will be observed. Whether the laws of physics for a contracting universe would be the same as those for the expanding phase is uncertain. Those who support the steady-state theory and the perfect cosmological principle are convinced that they would not, and that it is therefore idle to speculate about such a model, but not all cosmologists will go as far as this. They point out, as William Bonnor has done, that the laws of the behaviour of gases derived in the laboratory under restricted conditions of temperature and pressure, have been applied to explain the processes occurring inside stars where pressures and temperatures are vastly greater, yet this extrapolation has worked extremely well: in consequence, provided one is aware of what one is doing, there is no reason to reject at least the consideration of a contracting model. Some or all laws may require modifying, but until there is evidence to this effect we must proceed from what we know and the laws that appear to be valid in considerably different situations.

At its simplest, the oscillating universe performs just one oscillation. The universe is created with a big bang, expands, reaches a maximum size, and then begins to contract until it ends with another concentration of matter. On this simple picture it started as a singularity and ends as one. But cosmologists are now seriously considering a repeatedly oscillating universe, which will continually follow contraction by expansion, and expansion by contraction provided the density of matter in the universe is great enough to slow down the expansion, so that gravitational collapse can become the ruling factor. From observations of the red-shift of radio and optical galaxies, the rate of the present expansion phase of the universe may be determined, and from this it is possible to compute the average density of matter that must be present to cause expansion to cease: some seventeen billion tons per cubic light year. On the face of it, this figure seems high, for the density in large spiral galaxies like ours has been computed (from observation of their rotation rates and their stellar populations) and then averaged over the universe, to give a figure of only 600 million tons per cubic light year – a density some thirty times too small. But this assumes that galaxies provide most of the matter there is, and this may well be an incorrect assumption. There is evidence for the existence of intergalactic matter and for energy in the form of radiation and neutrinos. In relativity, energy possesses a mass equivalent, and so the radiation and neutrinos would add to the figure for the total mass, and hence to the density. However, further calculation indicates that the neutrino contribution appears to be too small – about 1000 times less than the average density computed taking masses of galaxies alone – and must be neglected. The familiar kinds of radiation also appear to make little if any contribution of a substantial size, and the discrepancy must be filled, if filled at all, by intergalactic matter.

Visual observation of clusters of galaxies provides a clue here, since it is found that the motions of members of a cluster show them to be held together more strongly than their masses alone would lead one to expect, and for large clusters the discrepancy is in fact thirty times. To investigate this matter which is not generally discernible at visual wavelengths, means further studies in invisible wavelengths, but already the latter have given the astronomer some useful evidence. From what has been described in Chapters 6, 7 and 8, it is clear that ideas about the evolution of stars and galaxies are having to be revised. The observation of X-ray stars confirms optical observations that the birth of stars appears still to be taking place and, moreover, that many may be formed as very close binaries that later move apart. Other X-ray observations indicate the probable presence of dying stars – novae – and of the possible presence of stars in the last stages of development, where they have shrunk to neutron star size, or even undergone or are close to the phenomenon of gravitational collapse. Indeed it is becoming increasingly clear that the Galaxy is a source of stars in every stage of development, with evidence for objects on and also far away from the main sequence – in other words a dynamic ever-changing stellar system. What is more, investigations of γ-rays and cosmic ray particles, and studies at radio wavelengths, have demonstrated the presence within the Galaxy of dust, gas and magnetic fields hitherto unsuspected: these have considerable implications for dynamic changes within the Galaxy.

Extra-galactic observations at invisible wavelengths have shown the existence of eruptive radio galaxies, and the fact that they display fluctuations of emission, while they have also drawn attention to the existence and possible importance of objects like the Seyfert galaxies. All this has made it clear that here, too, is a dynamic situation, and makes it seem more plausible to reject the gradual galactic evolutionary sequence that moved from irregular through spiral to elliptical forms, and see in every type, evidence of direct condensation, with the class of galaxy formed depending upon local conditions such as the speed of rotation of the generating gas cloud. Again, the radio astronomer has directed attention to the eruptions of great magnitude that are occurring in distant galaxies, and to the existence of the quasars. These present a particular problem: perhaps they are galaxies in the earliest stages of development – if the big bang theory is correct and if their distances are very great, then they indeed will be objects formed comparatively soon after the initial explosion. On the other hand, they can conceivably be large masses of material that are undergoing gravitational collapse, yet, if they are, it would seem likely that this must also mean that they are at the close of their lives – galaxies possessing little rotation that have shrunk under gravity after a life of radiation emission at a high rate? The solutions of this dynamic situation opened up by observations in the new wavelength ranges are manifold, but already some important investigations are beginning to show a way through the maze of possibilities. In particular the radio astronomers have some important fringe evidence, and this we must next consider.

CHAPTER 10

The New Problems

Of the theories of the universe that have been mentioned, none has raised so much controversy as the steady-state. To prove or disprove this hypothesis is important, since its removal will leave the field open to truly evolutionary models, while its confirmation will mean that such evolutionary change as there may be is somewhat restricted. A vital principle is at stake, and it is observation that must be the final arbiter. Visual observation seems unable to decide the matter – certainly it cannot do so alone – and it is the radio astronomer who has so far amassed the evidence that makes it seem probable that a decision is not far away.

In 1955 Martin Ryle and his colleagues at the Mullard Radio Astronomy Observatory, using their catalogued positions of radio sources, together with some results obtained from Australia, made the first observational attack. Ryle's analysis of his observations centred in a graph in which the number of radio sources in a given amount of sky was plotted against apparent radio brightness. If we ignore the motion of the sources, and hence any effect their red-shifts might have on their radio brightness, and if the sources are equally distributed in space – as the steady-state theory supposes – then calculation shows that, using a scale of the kind adopted by Ryle, the graph ought not only to be a straight line, but a straight line lying at an angle of 56° 18′ – a steep angle but one precisely calculable. Should the red-shifts be taken into account, then the curve will be a little less steep, at least for the more distant objects. On the other hand, if the big bang theory is correct, the more distant objects should appear in greater profusion in any given area of sky because we shall then be looking back into the universe at an earlier stage of its existence, before expansion has proceeded as far as it is seen to have done in the nearer reaches of space. In this case the curve will slope even more steeply.

The results Ryle found gave a slope of 72° 36′ – much too steep to support the steady-state hypothesis. However, further analysis of the observations and the use of more precise equipment, allowed Ryle to reduce the slope, so that by 1961 the evidence gave a result, also with a straight line on the graph but now with the slope reduced to 60° 57′, still considerably larger than the 56° 18′ required by the steady-state. Admittedly Ryle conceded that he had assumed all extra-galactic radio sources to possess the same intrinsic brightness, whereas those very far off would probably be intrinsically different since one observes them as they were, rather than as they are. This factor might make the curve have a slope that was less steep, but, on the other hand, it envisaged an evolution in radio sources, an evolution which according to the steady-state theory should not be noticeable since the universe appears the same at all times.

This may be countered by allowing a small evolutionary difference to be observable, assuming that what the radio astronomers have recorded is not a large enough sample of the universe to provide a homogeneous picture – a sample as big as our Galaxy obviously shows evolutionary differences among the stars, but this is clearly not large enough, so it is a very real question how large a satisfactory sample must be. What is more, as Dennis Sciama pointed out, it was not clear what the objects were that were being used in evidence. Were they radio galaxies or were they something else? Was there, perhaps, a confusion between two different kinds of source – were some radio galaxies but others a kind of radio star within the Galaxy?

Once quasars were discovered a new analysis could be made, and it then turned out that the radio galaxies did provide a slope of 56° 18′, but that quasars gave a very different result, with a slope of 66°. If quasars are very distant, then there is a true clustering effect – there are more quasars in distant parts of space than is to be expected if they were evenly distributed. And if quasars are so distant and therefore show extra-galactic objects in an early stage of development, this clustering is to be expected; the steady-state is wrong and the evolutionary form of the universe is correct, for the sample of space here involved is very great indeed. The one proviso is that it has not been established without question that the quasars are very distant. They could, Hoyle and G. Burbidge claim, be material ejected from our own Galaxy, an explanation that we have already seen is not beyond the bounds of possibility. Other galaxies are observed to undergo explosions and ejections and there is little reason to doubt that this could occur in our case.

Sciama also attacked the new results by considering the possibility that some quasars are extra-galactic and some within the Galaxy. He based his suggestion on the fact that red-shifts were known for some quasars but not for others, and he therefore sorted them out and laid those with no red-shift on one side, as it were, and concentrated his attention on the remainder. With a student, Martin Rees, yet another graph was drawn, this time plotting red-shifts against radio brightness. On the steady-state theory there should be the same number of 'red-shift' quasars between each set of values for red-shifts – from values from 0 to 1, 1 to 2, for instance – since in the expanding universe, red-shift is proportional to distance. But his results again show a greater number, a clustering, at large red-shifts and thus the universe is evolving and not in the steady-state. Of course, this analysis again depends upon the interpretation of the red-shifts as due to great distance, and the fact that some quasars give evidence of no such shift may be due to lack of detailed observation – after all, these objects are optically faint, and time for observing them with the 200-inch is limited since there are other important observing programmes that have a claim on its time. But the question remains undecided on the basis of quasar radio observations, especially those more recently made (1967 and 1968) at shorter radio wavelengths because these give a slope more favourable to the steady-state theory. But for anyone convinced that quasars are very distant, the steady-state theory seems in an uncomfortable position. Sciama has, indeed, now rejected the theory, but partly because of new observational evidence of a different kind.

Taking the big bang theory as correct, or even an oscillating universe as possible, an analysis shows that it seems likely that, close to the start of expansion, the temperature of the universe was very high. In particular, Robert Dicke who, like Hoyle and Narlikar, has been developing a new theory of gravitation that seems to be more in accordance with modern observation than either the theory proposed by Newton or that developed by Einstein, came to the conclusion that an oscillating universe is the most likely cosmological theory. And he felt certain that during the contracting phase before the present expansion, the temperature was high enough to burn out the heavy elements that had formed during the previous cycle, and reduce everything to hydrogen. For this the temperature required was not less than 10,000 million degrees, and although the universe is now expanding, Dicke and his colleagues at Princeton concluded that the remnant of this temperature should be detectable as a background radiation evenly distributed over the entire universe. The 'fireball', as Dicke calls it, with which the expansion phase commenced, cooled very rapidly indeed, the temperature dropping to 1000 million degrees within some four minutes when the diameter of the universe had reached seventy-eight light years, then in the next five to six minutes dropping another 900 million degrees, by which time expansion had increased the diameter ten times. By the end

of one month the temperature would be down to one million degrees
and the diameter up by a factor of 100. (The cooling is caused by the
radiation working against the gravitational pressure of the material
inwards.) Dicke and his colleagues felt so certain that this background
radiation should exist that in 1964 they prepared plans and began con-
structing a special radio telescope which would operate at 3·2 cm wavelength, a wavelength at which they reckoned the radiation should be
detectable at a reasonable strength.

While these preparations were being made, A. A. Penzias and R. W.
Wilson, two radio engineers at the Bell Telephone Laboratories who,
like Jansky before them, had been making observations of radio noise,
this time at a wavelength of 7·3 cm, reported to Dicke that they had
discovered a discrepancy in their results. The noise level they had
measured was higher than expected and consistently so, with the result
that they believed it to be caused by radiation received in their antenna,
radiation that was not directional in any way but appeared to be coming
from space. Later Wilson cooperated with P. G. Roll of Princeton and

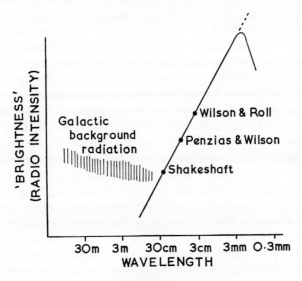

Figure 17

Observations of 'black body' radiation at various radio wavelengths plotted
against their intensity. Further observations, which must be made by space
probes, are required to show whether the radiation is a black body phenomenon (when it will follow the full line) or not (in which case it may follow the
dotted line).

the radiation was found on a wavelength of 3·2 cm. These were significant results, for calculation had shown that the radiation should have cooled to the low value of 3°K, and both the observations of Penzias and Wilson and of Wilson and Roll were consistent with this. However, if the radiation is a legacy of the fireball state of the universe at its early stages of expansion or previous contraction, it should be a black-body radiation – in other words its energy should be distributed in a specific way at different wavelengths. If hot like the Sun, it should radiate most strongly in the green part of the spectrum, but if as low as 3°K, its peak should lie at a wavelength of about 2½ mm, and a curve drawn of all the radiation – or rather that part which may be detected – is like that in figure 17.

Observations at a wavelength of 21 cm made by John Shakeshaft at the Mullard Radio Astronomy Observatory again gave results that not only confirmed that the radiation was detectable at this wavelength, but also that its intensity was that which was to be expected for a black body at 3°K. However, there are so far only three points on a straight line and although these fit the black body curve, they do not provide a unique result – the curve could, for instance, extend upwards to the right as shown by the dotted line in the figure. Nevertheless, in 1941, molecules of the substance cyanogen (CN) were discovered in the spectra of some late type stars, and a careful analysis made it clear that the bands of CN seemed due to something lying between the stars and the observer: the one trouble was that the cyanogen appeared to be excited by some other radiation which calculation showed must be at a temperature of 3°K. At the time, the result was not understood – there seemed to be no reason for the existence of such radiation and one notable astrophysicist suggested that the exciting temperature of 3°K was merely a conventional way of accounting for the effect, and was not to be taken literally! With the discovery of radio radiation measurements and the theoretical possibility of the existence of a background radiation of just this kind, the cyanogen observations become of notable importance. George Field of the University of California and Nick Woolf of the Goddard Institute for Space Studies, investigated the matter and found that to excite the cyanogen in the way observed, the radiation must have a specific intensity and possess this intensity at a wavelength of 2½ mm. When plotted on the curve of black-body radiation, the point fitted precisely, and is shown near the top of the curve.

Our Galaxy also provides a background radio radiation at wavelengths longer than 30 cm and, as the hatched line in figure 17 shows, this will mask the black-body radiation being investigated at the longer wavelengths, since the Galactic background is of greater intensity than theory shows the black-body radiation will be. Consequently it is only at wavelengths shorter than $2\frac{1}{2}$ mm that the background radiation can be effectively investigated further to determine whether or not it is truly of black-body type, and this means making observations from outside the atmosphere. Accordingly, arrangements are being made to do this and, if the observations provide intensity values that lie on the descending part of the curve on the right, there will be no doubt left that what is being observed is, in fact, black-body radiation, presumably from the early stages of the universe. This will then mean that strong observational evidence against the steady-state theory has been obtained, since it does not appear very likely that any modification of the theory can account for the existence of such universal radiation (if it is universal) at so low a temperature owing to primeval heating. Results from an Aerobee-Hi rocket indicate, however, that between 0.4mm and 1.3mm the temperature is about 8°K, too high for the black-body explanation.

Much will also depend upon the final decision about quasars. If they are systems undergoing gravitational collapse, or have been ejected from our Galaxy, then their significance in showing by their distribution that the steady-state theory is incorrect is vitiated. All the same the steady-state supporters have tried to incorporate Ryle's source counts by suggesting that the local part of space in which we find ourselves is a 'hole' in which galaxies are rather less densely distributed than anywhere else. M. S. Longair of the Mullard Observatory has concluded that this explanation fails, not for radio galaxies as such, but for quasars – in other words the hypothesis is not satisfactory for quasars if they have large distances. If the assumed quasar distances are wrong, there is then no need for the hypothesis. Another attack has been based on the argument that the synchrotron radiation causing quasar emission itself absorbs synchrotron emission at shorter wavelengths, since this would result in an apparent clustering due to the radio astronomer observing sources that appear too bright at his (longer) wavelengths and are thereby assumed to be nearer than they really are. On this explanation, the radio astronomer is seeing an even distribution because he is observing out further than he thinks. Longair finds that this argument is invalid when the radio radiation is compared at various wavelengths, the radio observations again failing to support the hypothesis.

When all is said and done, it must be admitted that the steady-state theory is receiving some hard knocks, and from some points of view it may appear unlikely that it can last very much longer. It has been valuable in the stimulus it has given to observers, as well as to cosmologists, and it is impossible to avoid the feeling that since the theory has shown surprising powers of resilience against what have, from time to time, appeared to be observational disproofs, it may once again surmount the difficulties that beset it.

Martin Ryle, however, believes that evidence points to the universe being evolutionary. He has recently suggested that quasars and radio galaxies are firm evidence of the evolution of galaxies, especially since he accepts the interpretation that the quasars are at vast distances, and therefore show the evolutionary universe far back in time. With Longair he has computed an evolutionary sequence in which the quasars, and indeed the radio galaxies, are considered as galaxies – probably elliptical ones – in the throes of formation. The radio radiation from these sources is considered to be generated by synchrotron radiation from a plasma within a magnetic field which, as will by now be clear, is a generally accepted idea, but in view of the radio evidence of multiple emissions from areas each side of the visual image, the plasma are thought of as ejected from the central mass. A possible evolutionary process is then worked out, and Ryle and Longair find that for not more than 1000 years the clouds may be expected to be no larger than the central gas cloud that generates the particularly blue optical emission. The energy of the plasma particles is here thought to be very high, and in due course they are ejected in two clouds in opposite directions at velocities that are very high indeed and comparable with the velocity of light. There now sets in a period, of the order of a hundred thousand years, in which the plasma clouds may be expected to pass through the galaxy's interstellar gas, and expand some ten thousand times. As they expand, electrified particles could be either continually accelerated or injected into the cloud, with the result that at a wavelength of 21 cm the radio brightness remains unchanging. But an analysis taking into account radiation at longer wavelengths such as eight metres, where there will be a change of radio brightness due to absorption within the cloud itself, shows that the constant emission at 21 cm is more likely to be due to an increase in the magnetic field in which the synchrotron radiation occurs. After a period of some ten thousand years, conditions will certainly change, since the increase in the magnetic field must reach a limit both because the clouds will move out into intergalactic space and because the plasma particles will have an equal amount of energy in every direction, so that increased interaction cannot occur: the plasma clouds will also continue to expand. The result of this is that the radio brightness will fall since there is now no mechanism to keep it constant as the volume of the clouds increases.

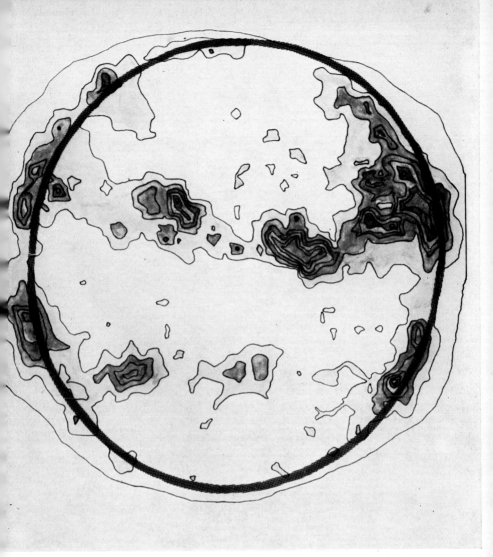

XI. The sun in ultraviolet. Map of the sun's ultraviolet activity at a particular time, observed by OSO-4 (plate IX). (The darker the area, the stronger the emission.) The areas emitting the very short wavelengths here recorded are estimated to be some 100,000 miles above the photosphere and are caused by emission of magnesium-10 ions at a temperature exceeding 825,000° C.

XII. The quasar 3C 273 shown very much enlarged from a photograph taken at Lick Observatory. The 'jet' of material is quite noticeable, and appears to be evidence for the vast release of energy from the object.

Thus, according to Ryle and Longair, the quasar stage and the radio galaxy stage are both temporary, and we are not observing here some permanent and peculiar object, as was first thought, but are watching the process of galactic evolution. This has not previously been observed optically since radiation in the visual wavelengths is too weak – due partly to the distribution of radiation at different wavelengths and partly to absorption in intergalactic space. But at radio wavelengths matters are different, the intergalactic absorption appears to be negligible, and the radio intensity is certainly greater than the visual, so the radio telescope has a greater power of space penetration.

Ryle and A. C. Neville have recently carried out a very detailed survey of a small area of sky centred round the north celestial pole and have found, as they expected, an increase in the number of sources per given area of sky as fainter and fainter objects are plotted. This has been a very sensitive survey, observing radio sources that are less intense than any previously plotted, and from an analysis of them Longair has put forward a most interesting and far-reaching idea. Down to a certain low intensity level the number of sources increases, but at a lower level still the number begins to decrease, so that at very great distances indeed there is a thinning out of sources. This is unlikely to be caused by observing at the 'edge' of the universe even though the distances are so great, and in any case the curvature of space would appear likely to prevent such an effect. As Ryle has suggested, it appears that in these results, which take the observer back in time to a period not so far from 10,000 million years ago, one is watching the universe in its very early stages, when there were few, if any, very strong radio sources, and when the condensation of galaxies was taking place.

This is indeed an astounding interpretation of an equally astounding series of observations, and if it leads us far from the steady-state theory it seems to bring us little nearer the solution of the question whether there was one big bang, or whether the primeval fireball was merely one along the endless succession of contractions, fireballs and expansions to be expected in an oscillating universe. There is much more work to be done by the cosmologist and theoretical astronomer and, above all, by the observer. The advent of radio astronomy has, in many ways, changed the picture of the universe, if not out of all recognition, at least beyond all previous conceptions, but it must be remembered that, except for a small range of 'visibility' in infra-red wavelengths and in the near ultra-violet, radio astronomy has had a start of many years over observations in the ultra short wavelengths. Yet it may well be that there are many surprises and much re-thinking to be done when more information, particularly in the X-ray and γ-ray ranges, is to hand. Already Sciama and Rees are analysing theoretically the absorption and other characteristics to be expected at these wavelengths in the radiation from plasma clouds of the kind associated with quasars, and it is obviously important that such observations must be conducted from rockets and satellites if the absorptions are to be checked, and more details of the fluctuation of emission at different wavelengths to be obtained, and if it is to be proved that quasars are, as Ryle believes, elliptical galaxies in the process of formation.

It would be wrong to leave the reader here lest he feel that there are now only details to be filled in, and that matters have reached a stage where the choice is a simple one between a big bang and an oscillating universe, that condensation into either spiral or elliptical galaxies is now established, or that current views on stellar evolution are the final word. The present situation is nowhere near as simple as it may appear.

Take, for instance, the abundance of helium in the universe. The general opinion has been that some helium was synthesized in the very earliest stages of the history of the universe, such synthesis occurring in the great heat of the primeval fireball of which the $3°K$ background radiation seems a confirmation. The older population II stars form with a little helium, and pour back into space helium and metals that they have synthesized, so that the population I stars that are formed later have atmospheres that are richer in metals and in helium than those of their predecessors. Yet some population I stars have been discovered with little or no helium, and this raises the question whether, in fact, there was a synthesis of helium early on. Should there not have been, then the primeval fireball cannot have been very hot, and difficulties therefore arise over the source of the $3°K$ background radiation. Possible escapes from the dilemma may be found by assuming that many neutrinos and anti-neutrinos were formed, with the result that helium synthesis would have been inhibited and there would have been a very short period of heat with a consequently short time for helium production, or perhaps material was unevenly distributed in primeval times so that there are now regions where little helium is to be found. It has even been suggested that the $3°K$ radiation is not of black-body type and is not universal, but is purely a local phenomenon in our own Galaxy and due perhaps to the absorption and re-radiation of starlight by interstellar grains: the steady-state theory could come into its own again if this were so. On the other hand, it may be that the observed cases of helium deficiency are not real, being caused either by too much reliance on direct spectroscopic evidence, or by the fact that the helium has sunk in a hydrogen-rich atmosphere and is thus rendered unobservable. But whatever the answer, it is clear that here is an important subject for investigation that may have far-reaching implications.

Again, the age of the universe seems rather too short for all galaxies to have formed from intergalactic gas, and this raises doubts about the big bang theory unless one takes Lemaître's interpretation, which is far from universally accepted. Moreover, Sciama and Rees claim to have found evidence for clustering of quasars, analogous to the clustering of ordinary galaxies and, if this is correct, it is likely that the volume of space we are observing can be no more than a comparatively small proportion of the whole. Some of the criticisms against the steady-state and arguments for a big bang may, in this case, have to be revised. Perhaps, too, it is rather too early to assess the full effects of new ideas of gravitation. Dicke has concentrated on the implications of his gravitational modifications as they effect the concept of an oscillatory universe, while Hoyle and Narlikar have asserted that all matter in the universe is the cause of the gravitational effects we observe, and this, again, has cosmological implications and effects on our ideas of physics that have yet to be completely determined. It also has implications in the realm of nuclear physics – one of the many points in the resuscitation of the steady-state cosmology that Hoyle is investigating. Another is the fact that matter and anti-matter may both be concentrated at the centre of galaxies, and thus give rise to immense releases of energy, perhaps of the kind to be observed in the elliptical galaxy M87 and in quasars. Again, this idea may make sense of the prodigious emission of infra-red radiation from the central regions of our own Galaxy, recently discovered by James Lequeux.

But Lequeux's work, as well as the investigations of British and South African observers, also underline the possibility that our knowledge of the distribution of interstellar dust will soon need to be revised, and we must recognize that the distribution found in our own Galaxy may not be copied in other galaxies of a similar kind. Perhaps, as de Vaucouleurs has pointed out, our present cosmologies may be in error because they tacitly suppose that our neighbourhood of space is typical of the whole universe, and it is of interest that Narlikar has suggested a 'bubble' universe, in which we may be considered to be living in an expanding bubble, that is only one of many such bubbles in a vaster cosmos than has previously been conceived.

It is clear, then, that we are far from being in a situation where only details need to be filled in: indeed it may be true to say that never before, in the whole history of astronomy, has there been a time so pregnant with new possibilities or so likely to witness a new revolution in the whole conceptual framework. But whatever the future may hold, one thing is certain, observations in the invisible wavelengths will provide the spearhead for the attack on the problems that astronomers and cosmologists now face.

Bibliographical Note

Since it takes time before the results of recent research find their way into books, readers who wish to keep up to date must consult the relevant literature. Most new results in invisible astronomy are announced in the weekly periodical *Nature*, or in the *Astrophysical Journal* and the *Monthly Notices of the Royal Astronomical Society*. These are highly technical journals, and the general reader may find it more profitable to consult the *New Scientist* (weekly), *Science* (monthly), *Science Journal* (monthly) and the *Scientific American* (monthly), *Sky and Telescope* (monthly), *Planetarium* (monthly).

The following books will also provide a background knowledge in greater detail than it has been possible to provide in this volume:

An Introduction to Astronomy by R. H. Baker and L. W. Frederick, van Nostrand, Princeton and London, 1968.
Essentials of Astronomy by Lloyd Motz and Annetta Duveen, Wadsworth, Belmont, Calif., 1966, Nelson, London, 1966.
Evolution of the Galaxies by V. C. Reddish, Oliver & Boyd, London, 1967.
Galaxies, Nuclei and Quasars by Fred Hoyle, Heinemann, London, 1965.
The Mystery of the Expanding Universe by W. Bonnor, Eyre & Spottiswoode, London, 1964, and Macmillan, New York, 1965.
Quasi-stellar Objects by E. and M. Burbidge, Freeman, San Francisco and London, 1967.
Radio Astronomy by F. Graham Smith, Penguin Books, Baltimore and London, 1960.
Space by Patrick Moore, Lutterworth Press, London, 1968.

Some readers may also wish to examine the earlier history of astronomy to obtain a fuller perspective of the latest research, and the following are recommended:

A History of Astronomy by A. Pannekoek, Allen and Unwin, London, 1961.
A History of Rocketry and Space Travel by W. von Braun and F. Ordway, Nelson, New Jersey and London, 1967.
Changing Views of the Universe by Colin A. Ronan, Eyre & Spottiswoode, London, 1961, and Macmillan, New York, 1962.

For specifically British contributions:

Their Majesties' Astronomers by Colin A. Ronan, Doubleday, New York, 1969, and Bodley Head, London, 1967.

Index